Applied Probability and Statistics (Continue

BROWNLEE · Statistical Theory an
  Engineering, *Second Edition*
CHAKRAVARTI, LAHA and RO
  Applied Statistics, Vol. I
CHAKRAVARTI, LAHA and ROY · Handbook of Methods of Applied Statistics, Vol. II
CHERNOFF and MOSES · Elementary Decision Theory
CHIANG · Introduction to Stochastic Processes in Biostatistics
CLELLAND, deCANI, BROWN, BURSK, and MURRAY · Basic Statistics with Business Applications
COCHRAN · Sampling Techniques, *Second Edition*
COCHRAN and COX · Experimental Designs, *Second Edition*
COX · Planning of Experiments
COX and MILLER · The Theory of Stochastic Processes
DANIEL and WOOD · Fitting Equations to Data
DAVID · Order Statistics
DEMING · Sample Design in Business Research
DODGE and ROMIG · Sampling Inspection Tables, *Second Edition*
DRAPER and SMITH · Applied Regression Analysis
ELANDT-JOHNSON · Probability Models and Statistical Methods in Genetics
GOLDBERGER · Econometric Theory
GUTTMAN, WILKS and HUNTER · Introductory Engineering Statistics, *Second Edition*
HAHN and SHAPIRO · Statistical Models in Engineering
HALD · Statistical Tables and Formulas
HALD · Statistical Theory with Engineering Applications
HOEL · Elementary Statistics, *Third Edition*
HUANG · Regression and Econometric Methods
JOHNSON and KOTZ · Distributions in Statistics
  Discrete Distributions
  Continuous Univariate Distributions-1
  Continuous Univariate Distributions-2
  Continuous Multivariate Distributions
JOHNSON and LEONE · Statistics and Experimental Design: In Engineering and the Physical Sciences, Volumes I and II
LANCASTER · The Chi Squared Distribution
LEWIS · Stochastic Point Processes
MILTON · Rank Order Probabilities: Two-Sample Normal Shift Alternatives
RAO and MITRA · Generalized Inverse of Matrices and Its Applications
SARD and WEINTRAUB · A Book of Splines
SARHAN and GREENBERG · Contributions to Order Statistics
SEAL · Stochastic Theory of a Risk Business
SEARLE · Linear Models
THOMAS · An Introduction to Applied Probability and Random Processes
WHITTLE · Optimization under Constraints
WILLIAMS · Regression Analysis

continued on back

Michael R. Cullen

Introduction to the Theory of Statistics

Introduction to the Theory of Statistics

Harold J. Larson

John Wiley & Sons, Inc.

New York London Sydney Toronto

Library of Congress Cataloging in Publication Data

Larson, Harold J. 1934–
Introduction to the theory of statistics.

(Wiley series in probability and mathematical statistics)
Bibliography: p. 229
1. Mathematical statistics. I. Title.

QA276.L315 519.5 72-5321
ISBN 0-471-51775-5

Printed in the United States of America

10 9 8 7 6 5 4 3 2 1

for
Douglas
Josef
Hugh
and
Rachel

Preface

This book is appropriate for a one-semester or two-quarter introduction to statistical theory for upper division undergraduate or first-year graduate students. It is assumed that the student has had a one-quarter or one-semester introduction to probability theory, at the level of Parzen (4) or Barr and Zehna (5); it is also assumed that he has had a course in linear algebra, from which he gained some facility with vectors and matrices.

Although this is a book on the theory of statistics, a large number of practical problems are discussed in the examples and exercises. I feel that discussions and derivations of theory alone, with no practical applications, may lead to great facility with integrals and other mathematical quantities, but these can be very sterile entities when divorced from applications. Since the application of statistics to real practical problems is not aways neat and straightforward, some of the examples discussed are a little complex and messy. I believe they are, however, typical enough to warrant the fairly lengthy discussion given.

Essentially all problems are phrased and attacked in the classical manner, the major exception being in Chapter 6 on Bayesian methods. I feel that the classical methods seem much more intuitive and understandable to the beginning student in statistics; indeed, in teaching decision theory and Bayesian methods, my experience has been that the students catch on much more quickly if they have had previous experience with classical techniques.

Chapter 1 is devoted to a rather rapid review of probability results that the student has previously been exposed to. The exception to this is a more complete discussion of vector random variables in Section 1.4, illustrated with the multinomial and multivariate normal random variables. Chapter 2 is devoted to populations, samples, and derivations of the standard sampling distributions.

Estimation of unknown parameters is discussed in Chapter 3, presenting the method of moments and maximum likelihood and properties of point estimates in some detail. Chapter 4 is devoted to a discussion of confidence sets, mostly interval estimates of single parameters, but also an introduction to confidence regions for vector parameters. Chapter 5 discusses the Neyman-Pearson theory of tests of hypotheses, some χ^2 tests and the sequential probability ratio test. Chapter 6 gives an introduction to decision theory and Bayesian methods and briefly applies this approach to estimation and tests of hypotheses. Chapter 7 is a fairly complete introduction to linear statistical models; this is the chapter which relies most heavily on the students' facility with vectors and matrices. Chapter 8 gives a short introduction to non-parametric procedures, based both on order statistics as well as on sample permutations. A total of 64 worked examples are scattered through the text, as are 350 exercises, ranging from quite mundane to extensions of the theory presented. Answers to the even numbered problems are presented in the Appendix. Also given in the Appendix are tables of the normal, χ^2, t, and F distributions.

Anyone who publishes a text on the theory of statistics owes a great debt to a large number of people who originally devised the techniques discussed. I am very grateful to all of them. I would also like to thank the Instituto de Matemática e Estatística da Universidade de São Paulo, the Fundaçao de Amparo a Pesquisa do Estado de São Paulo, and the Fulbright Commission of Brazil for partial support during a portion of the time needed to write this text. I would also like to thank Athena Shudde for an excellent job of typing a not-always-clear manuscript.

Monterey, California HAROLD J. LARSON

Contents

Introduction to the Theory of Statistics

CHAPTER 1

Probability

1.1 Introduction

Statistics has had a great influence on most, if not all, modern scientific disciplines. Most of the sciences have grown and developed by proposing various theories about one or more aspects of the real world. In checking whether the proposed theory is valid or not, statistical methods are very frequently used. Indeed, in any situation requiring an inference, in which it is not possible to construct a proof or disproof of a theory using the laws of mathematical logic, statistical reasoning is necessary. In such cases, it is generally possible to gather evidence in support of or in contradiction to the proposed theory; however, this evidence generally is not equivalent to a mathematical proof. It may still be possible that the theory is correct or incorrect, regardless of the evidence. However, if statistical reasoning and probability theory can be used to state that the theory is highly unlikely, in the face of the observed evidence, then the theory should likely be discarded and usually is by rational people.

Through the years, many different statistical methods and procedures have been devised, to serve specific needs that have arisen in various fields. This book describes the theory behind many of the most commonly used methods and illustrates the theory with examples. It is assumed that the reader has had a previous course in probability theory, at the level of Parzen (4) or Barr and Zehna (5); the remainder of Chapter 1 summarizes the necessary probability concepts and reviews the most commonly used distributions of random variables. It is also assumed that the reader has had a course in linear algebra and matrix theory; these topics are especially useful in the description of vector random variables and in the general linear hypothesis theory. A course at the level of Perlis (1) or Stein (3) should prove sufficient for such needs. Since the reader may not have had a great deal of experience with the multivariate normal distribution, some of the useful results concerning it are derived in this chapter.

1.2 Probability Functions and Random Variables

An *experiment* is any operation whose outcome cannot be predicted in advance, with certainty. The *sample space* for an experiment is the set of all possible outcomes that might be observed upon completion of the experiment; frequently, more than one sample space may be defined for the same experiment. An *event* is a subset of the sample space; if the sample space is discrete, then every subset is an event. If the sample space is continuous, events are Borel sets; in any practical example we shall consider, any subset of interest is a Borel set and thus is an event. A *probability function* P is a rule which associates a real number $P(E)$ with every event E such that the following three axioms are satisfied:

1. $P(S) = 1$ where S is the sample space;
2. $P(A) \geq 0$ where A is any event, $A \subset S$.
3. $P(A_1 \cup A_2 \cup \cdots) = P(A_1) + P(A_2) + \cdots$ if $A_i \cap A_j = \phi$ for all $i \neq j$.

As a result of these axioms, it can be shown that $P(\phi) = 0$, $P(\bar{A}) = 1 - P(A)$, where $\bar{A}$ is the complement of A, and $P(A \cup B) = P(A) + P(B) - P(A \cap B)$ for any two events A and B.

The *conditional probability* of an event A, given that a second event B has occurred, is $P(A \mid B) = P(A \cap B)/P(B)$, where $P(B) > 0$. Note, then, that $P(A \cap B)$ may be computed from $P(B)P(A \mid B)$ or $P(A)P(B \mid A)$. If events $E_1, E_2, \ldots, E_n$ are *mutually exclusive* ($E_i \cap E_j = \phi$ for all $i \neq j$) and $E_1 \cup E_2 \cup \cdots \cup E_n = S$, then $E_1, E_2, \ldots, E_n$ form a *partition* of S. Given that $E_1, E_2, \ldots, E_n$ form a partition of S, then for any event $A \subset S$,

$$P(E_i \mid A) = \frac{P(A \mid E_i)P(E_i)}{\sum_{i=1}^{n} P(A \mid E_i)P(E_i)}, \qquad \text{(Bayes' Theorem)}$$

Example 1.2.1

Assume that the probability is .95 that the jury selected to try a criminal case will arrive at the appropriate verdict. That is, given a guilty defendant on trial, the probability is .95 that the jury will find him guilty and, conversely, given an innocent man on trial, the probability is .95 that the jury will find him innocent. Suppose that the local police force is quite diligent in its duties and that 99% of the people brought before the court are actually guilty. Let us compute the probability that a defendant is innocent, given that the jury finds him innocent. Let G be the event that the defendant is guilty and let J be the event that the jury finds him guilty. Then, we are given that $P(J \mid G) = P(\bar{J} \mid \bar{G}) = .95$, $P(G) = .99$, and we want to compute

$P(\bar{G} \mid \bar{J})$. G and $\bar{G}$ form a partition of the sample space, so from Bayes' Theorem

$$P(\bar{G} \mid \bar{J}) = \frac{P(\bar{J} \mid \bar{G})P(\bar{G})}{P(\bar{J} \mid \bar{G})P(\bar{G}) + P(\bar{J} \mid G)P(G)}$$

$$= \frac{(.95)(.01)}{(.95)(.01) + (.05)(.99)} = .161.$$

Thus, there is about 1 chance in 6 he really is innocent, if found innocent, and about 5 chances in 6 he is guilty, although found innocent. Similarly, the probability is .0005 that he is innocent, if found guilty, and is .9995 that he is in fact guilty if found guilty.

Two events, A and B, are *independent* if and only if $P(A \cap B) = P(A)P(B)$. It follows then, that if A and B are independent,

$$P(A \mid B) = P(A) \qquad \text{and} \qquad P(B \mid A) = P(B).$$

Thus, the conditional probability of an event occurring is the same as the unconditional, if it is independent of the conditioning event. The n events, $A_1, A_2, \ldots, A_n$ are completely independent if and only if

$$P(A_i \cap A_j) = P(A_i)P(A_j) \qquad \text{for all } i \neq j,$$

$$P(A_i \cap A_j \cap A_k) = P(A_i)P(A_j)P(A_k) \qquad \text{for all } i \neq j \neq k$$

$$\cdots\cdots\cdots\cdots\cdots\cdots\cdots\cdots\cdots$$

$$P(A_1 \cap A_2 \cap \cdots \cap A_n) = \prod_{i=1}^{n} P(A_i).$$

Example 1.2.2

A major use of this definition of independence is the assignment of probabilities to the *single-element events* for an experiment consisting of independent trials. Suppose, for example, that a new Chevrolet is equipped with 5 new tires, selected at random from a supply of new tires, and that the probability of a single tire from this source being defective (in one way or another) is .01. Since the tires are selected in a haphazard way, for this car, it would seem reasonable that the occurrence or nonoccurrence of defects from one tire to the next are independent. Letting n stand for nondefect and d for defect, a reasonable sample space for the experiment of selecting 5 tires is

$$S = \{(x_1, x_2, \ldots, x_5)\colon\ x_i = n \text{ or } d,\ i = 1, 2, \ldots, 5\}.$$

Thus, S has $2^5 = 32$ elements and probabilities can easily be assigned to the single element events (5-tuples) by multiplication. Thus, for example,

$$P((n, n, n, n, n)) = (.99)^5$$
$$P((d, d, d, d, d)) = (.01)^5$$
$$P((n, d, d, d, d)) = (.99)(.01)^4;$$

the probability of exactly one defective tire in the 5 selected then is

$$\binom{5}{1}(.99)^4(.01) = .048$$

and the probability of no defective tires in the 5 selected is $(.99)^5 = .951$.

Random variables are very basic in developing most statistical methods. A *random variable* is simply a real-valued function of the elements of a sample space. The probability function for the sample space can then be used to make probability statements about the value the random variable may attain when the experiment is performed; since the range of the random variable is necessarily a subset of the real line, many authors say that the random variable *induces* a probability measure on the real line.

A random variable may be either *discrete* or *continuous*, depending on its range. In either case, the *distribution function* for the random variable provides a description of its behavior; the distribution function for a random variable X is defined as follows:

$$F_X(t) = P(X \leq t) \qquad \text{for all real } t.$$

Note, then, that

$$P(a < X \leq b) = F_X(b) - F_X(a).$$

If X is *discrete*, its *probability function* $p_X(t)$ can also be used to describe its behavior, where

$$p_X(t) = P(X = t) = F_X(t) - F_X(t-);$$

by definition

$$F_X(t-) = \lim_{h \to 0} F_X(t - h), \qquad \text{where } h > 0.$$

The range R_X of a discrete random variable has as elements all real numbers t such that $p_X(t) > 0$. This is necessarily a discrete set and thus the random variable is called discrete.

Example 1.2.3

Chuck-a-luck is a popular gambling game at the casinos in Las Vegas. The game is played as follows: The croupier has 3 fair dice; a play consists of one roll of the 3 dice. You, as the better, can bet, say, \$1 on the occurrence of any of the integers 1 through 6; suppose you bet on the occurrence of 3's.

Then, when the dice are rolled, if one 3 occurs you win \$1, if two 3's occur you win \$2 and if three 3's occur, you win \$3. If no 3's occur you lose your \$1. Let X be the amount you will win on one play of this game. Then $R_X = \{-1, 1, 2, 3\}$ and the probability function for X is

$$\begin{aligned} p_X(x) &= \frac{125}{216} && \text{for } x = -1 \\ &= \frac{75}{216} && \text{for } x = 1 \\ &= \frac{15}{216} && \text{for } x = 2 \\ &= \frac{1}{216} && \text{for } x = 3. \end{aligned}$$

If X is a *continuous* random variable then its distribution function F_X is continuous and

$$\lim_{h \to 0} F_X(t - h) = F_X(t),$$

for all t. Thus its probability function is identically zero and would be of no use in computing probabilities for X. However, the slope of F_X at t gives the relative rate at which probability is accumulating at t; thus the *density function* for a continuous random variable is

$$f_X(t) = \frac{d}{dt} F_X(t)$$

at all points where the derivative exists. Then,

$$\begin{aligned} P(a < X \leq b) &= F_X(b) - F_X(a) \\ &= \int_a^b f_X(t)\, dt \end{aligned}$$

and integration of the density function enables us to evaluate the probability that X lies in the interval (a, b). The range, R_X, of a continuous random variable is the set of real numbers t such that $f_X(t) > 0$.

Example 1.2.4

A traffic light at a major intersection, in the direction you travel, is green 2 minutes, red 1 minute, then green again, etc., with the green light starting on the hour every hour. On your way to work in the morning you are equally likely to arrive at this intersection at any instant between 8:05 a.m.

and 8:15 a.m. If we let X be the number of minutes past 8:05 at which you arrive at the intersection, then X is a continuous random variable, and thus has a density function f_X. The fact that you are equally likely to arrive at any instant between 8:05 and 8:15 a.m. means that the density function for X is constant over this interval. Thus we require $f_X(x) = C$, $0 < x < 10$. Since the integral of the density over the range of X must be 1, $C = (1/10)$ and we have

$$f_X(x) = \frac{1}{10}, \qquad 0 < x < 10.$$

If you arrive between 8:06 and 8:08, or between 8:09 and 8:11, or between 8:12 and 8:14, then the light is green. Thus, the probability that the light is green when you arrive is

$$\int_1^3 \frac{1}{10}\,dx + \int_4^6 \frac{1}{10}\,dx + \int_7^9 \frac{1}{10}\,dx = 3\left(\frac{2}{10}\right) = .6.$$

The rth *moment of a random variable* X, $r = 1, 2, 3, \ldots$ is $E[X^r]$, the expected value of X^r; if X is discrete

$$E[X^r] = \sum_{x \in R_X} x^r p_X(x),$$

whereas if X is continuous

$$E[X^r] = \int_{R_X} x^r f_X(x)\,dx.$$

The first 2 moments are the most frequently used in describing random variables. The *mean* is $\mu_X = E[X]$ and the *variance* is $\sigma_X^2 = E[(X - \mu_X)^2]$, the *standard deviation* is $\sigma_X = \sqrt{\sigma_X^2}$. For most of the standard distributions, the moment generating function, $m_X(t)$, exists and provides a useful way of evaluating moments. By definition,

$$m_X(t) = \sum_{x \in R_X} e^{tx} p_X(x) \qquad \text{if } X \text{ is discrete}$$

$$= \int_{R_X} e^{tx} f_X(x)\,dx \qquad \text{if } X \text{ is continuous.}$$

Then, $m_X^{(r)}(0) = E[X^r]$, if m_X exists in a neighborhood of $t = 0$.

The *factorial moment generating function* (also called the *probability generating function*) is

$$\xi_X(t) = E[t^X] = m_X(\ln t).$$

It is easily verified that

$$\xi_X^{(r)}(1) = E[X(X-1)\cdots(X-r+1)]$$

and that $\xi_X^{(r)}(0) = r!p_X(r)$. The *cumulant generating function* is

$$c_X(t) = \ln m_X(t);$$

note that

$$c_X^{(1)}(0) = \mu_X$$

$$c_X^{(2)}(0) = \sigma_X^2.$$

EXERCISE 1.2

1. Prove, from the axioms, that

$$P(A \cup B) = P(A) + P(B) - P(A \cap B)$$

for any events A and B.

2. Prove, from the axioms, that a probability function must be monotonic; that is, if $A \subset B$, then $P(A) \leq P(B)$.

3. Given $S = \{1, 2, 3\}$, $A = \{1\}$, $B = \{3\}$, $C = \{2\}$, $P(A) = 1/2$, $P(B) = 1/5$, find

(a) $P(C)$ (d) $P(\bar{A} \cap B)$

(b) $P(A \cup B)$ (e) $P(\bar{A} \cup \bar{B})$

(c) $P(\bar{A})$ (f) $P(B \cup C)$

4. A pair of fair dice is rolled one time. Let X be the sum of the two numbers that occur. Compute p_X.

5. If two fair dice are rolled one time, what is the probability that both show the same face?

6. A die is loaded so that the probability of face i occurring is proportional to i, $i = 1, 2, \ldots, 6$. What is the probability of an even number occurring when the die is rolled?

7. In a certain city, 40% of the registered voters are Republicans, 50% are Democrats, and the remainder are Independent. Seventy percent of the Republicans, 90% of the Democrats, and 50% of the Independents favor a school bond issue. If a registered voter is selected at random, what is the probability he favors the school bond issue? A person was selected at random, and he did favor the bond issue. What is the probability that he is registered as an Independent? as a Democrat?

8. Sixteen teams are entered in a basketball tournament. No game can end in a tie and any team which loses a game is eliminated. Thus a total of 4 rounds is required to determine the champion. Assume that your favorite team has probability .9 of winning its first game and conditional probabilities of .8, .7, and .6 of winning its succeeding games, respectively, given that it won those preceding. What is the probability that your team is eliminated in the third round? that it wins the tournament?

9. A fair coin is flipped two times. Let A be the event that a head occurs on the first flip, B that a head occurs on the second flip, and C that the two faces match. Are A, B, and C independent?

10. A student is given a true–false exam with 10 questions. If he gets 8 or more correct he passes the exam. Given that he guesses at the answer to each question, compute the probability that he passes the exam.

11. A ten-inch piece of string is cut into two pieces at a random point along its length. What is the probability that the longer piece is at least twice as long as the shorter piece?

12. A rifle is fired at a target until a bullseye is scored. If the probability of a bullseye is .9 for each shot, what is the probability that an odd number of shots are required to score the first hit?

13. A student is presented with 5 word pairs and allowed as much time as he likes to memorize them. Twenty-four hours after he is done he is presented with the first words of each pair and asked to recall the associated words. Assuming that he has the same probability p of recalling each associated pair, and that the recollections are independent from one pair to the next, what is the probability function for X, the number he is able to correctly recall?

14. If in question 13, the probability of correctly recalling association i is p_i, $i = 1, 2, \ldots, 5$, what is the probability function for X?

15. If the expected values exist, show that $E[X^2] \geq \{E[X]\}^2$.

16. Show that $E[(X - b)^2]$ is minimized by taking $b = \mu_X$.

17. The quantities $E[(X - \mu_X)^r]$ are called the *central moments* of X; show that the derivatives of $m_{X-\mu}(t) = e^{-\mu t} m_X(t)$ at $t = 0$ generate the central moments.

18. Verify that the first two derivatives of $c_X(t) = \ln m_X(t)$, at $t = 0$, give μ_X and σ_X^2.

19. Verify that, if X has $R_X = \{x: \ x = 0, 1, \ldots, n\}$, then

$$p_X(k) = \frac{\xi_X^{(k)}(0)}{k!}, \qquad k = 0, 1, 2, \ldots, n.$$

20. If X has density

$$f_X(x) = e^{-x}, \qquad x > 0$$

find its distribution function.

21. A discrete random variable Y has distribution function

$$F_Y(t) = 0 \qquad t < 0$$
$$= 1 - q^{[t]} \qquad t \geq 0$$

where $0 < q < 1$, $[t]$ is the integer part of t; i.e., $[t] = i$ for $i \leq t < i + 1$, $i = 0, 1, 2, \ldots$. What is the probability function for Y?

22. Let $\eta_1, \eta_2, \ldots$ denote the moments of a random variable X (thus $\eta_k = m_X^{(k)}(0)$) and let $\xi_1, \xi_2, \ldots$ denote the factorial moments of the same random variable ($\xi_k = \xi_X^{(k)}(1)$). Express the first 3 moments in terms of the first 3 factorial moments and vice versa.

1.3 Some Standard Distributions

The simplest, standard discrete random variables are defined in experiments consisting of *Bernoulli trials*, trials which have only two possible outcomes, success (s) or failure (f). The *Bernoulli random variable* X is the number of successes observed in a single Bernoulli trial; its range is $R_X = \{0, 1\}$ and its probability function is

$$p_X(x) = p^x(1 - p)^{1-x}, \qquad x = 0, 1, \qquad 0 \leq p \leq 1,$$

where p is the probability of success on the Bernoulli trial. For the Bernoulli random variable, $m_X(t) = q + pe^t$, $\mu_X = p$, $\sigma_X^2 = pq$, where $q = 1 - p$.

The *binomial random variable* X is the number of successes to occur in an experiment which consists of n repeated, independent Bernoulli trials, where again the probability of success on any trial is p. Its range then is $R_X = \{0, 1, 2, \ldots, n\}$ and its probability function is

$$p_X(x) = \binom{n}{x} p^x(1 - p)^{n-x}, \qquad x \in R_X, \qquad n = 1, 2, 3, \ldots, 0 \leq p \leq 1.$$

Its moment generating function is $m_X(t) = (q + pe^t)^n$, $\mu_X = np$, $\sigma_X^2 = npq$. Note that the binomial random variable is the (random) number of successes to be observed in a fixed number n of independent Bernoulli trials and, if $n = 1$, X is Bernoulli.

Now suppose that repeated, independent Bernoulli trials are performed until a fixed number, r, of successes has occurred. The (random) number of trials required, X, is called the *negative binomial random variable*. Its range is the infinite set $R_X = \{r, r + 1, r + 2, \ldots\}$ and its probability function is

$$p_X(x) = \binom{x-1}{r-1} p^r q^{x-r} = \binom{x-1}{x-r} p^r q^{x-r}, \qquad x \in R_X,$$

$$r = 1, 2, 3, \ldots, 0 < p \leq 1.$$

Its factorial moment generating function is $\xi_X(t) = (pt)^r/(1 - qt)^r$, $\mu_X = r/p$, $\sigma_X^2 = rq/p^2$. If the number of successes required is $r = 1$, then X is called the *geometric random variable;* its range is, of course, $R_X = \{1, 2, 3, \ldots\}$, and its probability function is

$$p_X(x) = pq^{x-1}, \qquad x \in R_X, \qquad 0 < p \leq 1;$$

its factorial moment generating function is $\xi_X(t) = pt/(1 - qt)$, $\mu_X = 1/p$, $\sigma_X^2 = q/p^2$. (Some authors call $Y = X - 1$, the number of trials *before* the first success, a geometric random variable; note then that $R_Y = \{0, 1, 2, \ldots\}$ and

$$p_Y(y) = pq^y, \qquad y = 0, 1, 2, \ldots .)$$

An urn contains N balls, of which M are white; n balls are drawn at random, without replacement, from the urn. Let X be the number of white balls in the n that are drawn. X is called the *hypergeometric random variable;* its range is $R_X = \{x = \max(0, M - n), x + 1, x + 2, \ldots, \min(n, M)\}$, and its probability function is

$$p_X(x) = \frac{\binom{M}{x}\binom{N - M}{n - x}}{\binom{N}{n}}, \qquad x \in R_X.$$

N and n are positive integers, M is a nonnegative integer. The moment generating function for X is not an easily expressed function; the mean of X is $\mu_X = n(M/N)$ and

$$\sigma_X^2 = n\frac{M}{N}\left(1 - \frac{M}{N}\right)\frac{N - n}{N - 1}.$$

Note that if the balls had been drawn with replacement, rather than without, then X would be binomial with $p = M/N$.

A *Poisson process* is a process generating occurrences (called events) in a continuum (of time, distance, area, volume, etc.) such that

(a) It is possible to take a sufficiently short interval, of length h, in the continuum such that
   (i) The occurrence of exactly one event in the interval of length h is approximately λh (proportional to h) no matter where this interval is located in the continuum.
   (ii) The occurrence of two or more events in this interval of length h is approximately 0.

(b) The occurrences of events in nonoverlapping intervals of length h are independent.

(The constant of proportionality, λ, is called the *rate parameter* of the process.) Many processes in the real world seem to adequately satisfy these assumptions; examples are the emission of electrons from a radioactive source, locations of nonclustering bacteria in a volume of water, arrival times of autos at a parking lot, or of orders for a certain item from the stock of a supplier. There are several random variables which can be defined on a Poisson process.

The *Poisson random variable* X is defined on a Poisson process with rate parameter λ; it is the (random) number of events to occur in an interval of fixed length s. Its range is $R_X = \{0, 1, 2, 3, \ldots\}$ and its probability function is

$$p_X(x) = \frac{(\lambda s)^x}{x!} e^{-\lambda s}, \qquad x \in R_X, \qquad \lambda > 0, \qquad s > 0;$$

its moment generating function is $m_X(t) = e^{\lambda s(e^t - 1)}$, $\mu_X = \lambda s$, $\sigma_X^2 = \lambda s$.

A *uniform random variable* X on the interval (a, b) (which we shall refer to as X is uniform, (a, b)) has range $R_X = \{x\colon\ a < x < b\}$ and density function $f_X(x) = 1/(b - a)$, $X \in R_X$, a and b are real, $a < b$; its moment generating function is $m_X(t) = (e^{tb} - e^{ta})/t(b - a)$, $\mu_X = (a + b)/2$, $\sigma_X^2 = (b - a)^2/12$. Since its density function is constant on the interval (a, b), we shall frequently say that X is equally likely to equal any of the numbers in the interval (a, b); we really mean by this that the probability X falls in any interval contained in (a, b) is proportional to the length of that interval.

The *gamma random variable* X can also be defined on a Poisson process with rate parameter λ; it is the continuous amount of time (or area, distance, etc.) required to observe the rth event, starting from an arbitrary point in the process; r is fixed. Its range is $R_X = \{x; x > 0\}$ and its density function is

$$f_X(x) = \lambda \frac{(\lambda x)^{r-1}}{\Gamma(r)} e^{-\lambda x}, \qquad x \in R_X, \qquad r > 0, \qquad \lambda > 0;$$

its moment generating function is $m_X(t) = 1/[1 - (t/\lambda)]^r$, $\mu_X = r/\lambda$, $\sigma_X^2 = r/\lambda^2$. As derived from a Poisson process, the parameter r would have to be a positive integer; however, if the need arises, the above density can be used with r any positive number, integral or not. A very important special case of the gamma density is $r = 1$; this gives the *exponential random variable* with parameter λ. Thus, if X is an exponential random variable with parameter λ, its range is $R_X = \{x\colon\ x > 0\}$ and its density function is

$$f_X(x) = \lambda e^{-\lambda x}, \qquad x \in R_X, \qquad \lambda > 0;$$

its moment generating function is $m_X(t) = 1/[1 - (t/\lambda)]$, $\mu_X = 1/\lambda$, $\sigma_X^2 = 1/\lambda^2$. One way in which the exponential random variable occurs is as the length of time until the first occurrence in a Poisson process. Another important special case of a gamma random variable is given by $\lambda = 1/2$, $r = n/2$, where n is a positive integer; this is called the χ^2 *random variable* with n degrees of freedom. We shall see how it arises naturally in certain applications in Chapter 2.

The *beta random variable* X has range $R_X = \{x: \quad 0 < x < 1\}$ and density

$$f_X(x) = \frac{\Gamma(\alpha + \beta)}{\Gamma(\alpha)\Gamma(\beta)} x^{\alpha-1}(1 - x)^{\beta-1}, \qquad x \in R_X, \quad \alpha > 0, \quad \beta > 0;$$

the moment generating function is not very tractable, but the moments themselves are easily derived directly; it can be verified that $\mu_X = \alpha/(\alpha + \beta)$, $\sigma_X^2 = \alpha\beta/[(\alpha + \beta)^2(\alpha + \beta + 1)]$. In Chapter 2 we shall derive the distribution of an *F random variable* and of a *t random variable*. As we shall see, $X = (n/m)F/[1 + (n/m)F]$ has the beta distribution with $\alpha = n/2$, $\beta = m/2$ and thus the beta random variable is a transform of the F; similarly, we shall find that the square of a t random variable is an F random variable, and thus the beta distribution is also a transform of the t distribution. Note as well that if $\alpha = \beta = 1$, then X is uniform on (0, 1). In studying Bayesian techniques the beta distribution plays an important role as a *prior distribution*.

The *normal random variable* X has range $R_X = \{x: \quad -\infty < x < \infty\}$ and density function

$$f_X(x) = \frac{1}{\sigma\sqrt{2\pi}} e^{-(x-\mu)^2/2\sigma^2}, \qquad -\infty < \mu < \infty, \qquad \sigma > 0;$$

its moment generating function is $m_X(t) = e^{t\mu+(t^2/2)\sigma^2}$ and $\mu_X = \mu$, $\sigma_X^2 = \sigma^2$. Thus, the two parameters μ and σ in the normal density are the mean and standard deviation of X, respectively. The normal density frequently occurs in practical problems, both because it occurs naturally in many cases and because of the central limit theorem (see section 1.4). We shall use the shorthand notation $X \sim N(\mu, \sigma)$ to mean X is a normal random variable with mean μ and standard deviation σ. If $\mu = 0$, $\sigma = 1$, then X has the standard normal distribution, whose distribution function is given in Table 1. We shall define z_k by $P(Z \leq z_k) = k$, where Z is $N(0, 1)$. Because of the symmetry of the $N(0, 1)$ density, $z_{1-k} = -z_k$, for $0 < k < 1$.

EXERCISE 1.3

1. There is a constant probability p that an item produced by an assembly line will have a certain defect, and the occurrences of defects are independent. If 10 items are selected from the output of this assembly line, what is the probability function for X, the number of defects in the sample selected?

2. Assume that fatal auto accidents on the highways of the United States occur like events in a Poisson process with rate 2 per hour. What is the probability function for the number of fatal accidents to occur in a 24-hour period? What is the expected number of fatal accidents in such a period?

3. In a small town, there are 100 adults, 60 of whom are smokers. A random sample of 3 persons is selected from these adults; let Y be the number of smokers in the sample and give the probability function for Y. What is the probability that a majority of people in the sample are smokers?

4. The instants at which a used-car salesman makes a sale occur like events in a Poisson process with a rate of 1 per week. Starting with the beginning of work on Monday, what is the expected length of time until he makes his first sale? his second sale?

5. Henry drives to work along the same route every day. Assume that the probability he has an accident is p for each day and that the occurrences of his accidents are independent. What is the expected number of days until his first accident? his second accident?

6. Derive the distribution function for a geometric random variable X with parameter p.

7. Verify that the mean of a beta random variable with parameters α and β is $\alpha/(\alpha + \beta)$ and that its variance is $\alpha\beta/[(\alpha + \beta)^2(\alpha + \beta + 1)]$. Also verify that the maximum of its density function occurs at $(\alpha - 1)/(\alpha + \beta - 2)$ (when $\alpha + \beta \neq 2$). The value of x which maximizes f_X is called the *mode* of X.

8. Find the mode of a gamma random variable and verify the values given for its mean and variance.

9. Find the mode of a normal random variable.

10. The *median* of a continuous random variable is the value m_e such that

$$P(X \leq m_e) = P(X \geq m_e) = .5.$$

Find the median of the exponential random variable with parameter λ and the median of a normal random variable with parameters μ and σ.

11. Under what conditions are the mean and the median of a beta random variable equal?

12. Is it possible to choose the parameters of a gamma random variable so that its mean and median are equal?

13. Using repeated integration by parts, verify that

$$\int_0^t \lambda \frac{(\lambda x)^{r-1}}{\Gamma(r)} e^{-\lambda x}\, dx = 1 - e^{-\lambda t} \sum_{j=0}^{r-1} \frac{(\lambda t)^j}{j!}\,.$$

Thus, the distribution function with argument t of a gamma random variable X with parameters r (integer) and λ, $F_X(t; r, \lambda)$, is equal to one minus the distribution function F_Y of a Poisson random variable Y, at $r - 1$, with parameter λt; that is

$$F_X(t; r, \lambda) = 1 - F_Y(r - 1; \lambda t)$$

and a table of the Poisson distribution function can also be used to evaluate specific values of gamma distribution functions, and vice versa. Since both these random variables can be defined on the same Poisson process, it is also instructive to note that this equation says that the event: {time to the rth occurrence is less than or equal to t} is equivalent to the complement of the event: {there are $r - 1$ or fewer occurrences in the interval $(0, t)$}.

14. Using repeated integration by parts, show that

$$\int_0^p \frac{\Gamma(n+1)}{\Gamma(k)\Gamma(n-k+1)} x^{k-1}(1-x)^{n-k}\,dx = \sum_{i=k}^{n} \binom{n}{i} p^i (1-p)^{n-i}.$$

Thus, the distribution function of a beta random variable X with parameters k and $n - k + 1$, evaluated at p, $F_X(p; k, n - k + 1)$, is equal to one minus the distribution function of a binomial random variable Y at $k - 1$, with parameters n and p; i.e.,

$$F_X(p; k, n - k + 1) = 1 - F_Y(k - 1; n, p)$$

and a table of the binomial distribution function can also be used to evaluate certain values of beta distribution functions and vice versa.

15. If X is a continuous random variable with distribution function F_X, show that $Y = F_X(X)$ is uniform (0, 1). Conversely, if Y is uniform (0, 1), and F_X is any monotonic nondecreasing continuous function of a real variable with range (0, 1) (and F_X^{-1} is the inverse function) show that $F_X^{-1}(Y)$ has distribution function F_X. This is called the *probability integral transform* and is frequently used in computer routines to generate observed values of continuous random variables with arbitrary distribution from uniform (0, 1) random variables.

16. If X has distribution function F_X and $Y = a + bX$, $b > 0$, show that

$$F_Y(t) = F_X\left(\frac{t-a}{b}\right).$$

17. If X has continuous distribution function F_X and $Y = a + bX^2$, $b > 0$, show that

$$F_Y(t) = F_X\left(\sqrt{\frac{t-a}{b}}\right) - F_X\left(-\sqrt{\frac{t-a}{b}}\right)$$

where $t > a$.

18. In 17, if X is continuous so is Y; find the density of Y in terms of f_X.

19. Suppose X is standard normal, i.e., $\mu = 0$, $\sigma = 1$; find the density and distribution function for $Y = X^2$. (As we shall see in Chapter 2, Y is called a χ^2 random variable with 1 degree of freedom.)

20. The 100pth *percentile*, ξ_p, of a continuous random variable X is defined by

$$P(X \leq \xi_p) = p \qquad 0 < p < 1.$$

Find the percentiles of a uniform random variable on (0, 1); on (a, b).

21. Using the result in problem 19, show how tables of the standard normal distribution can be used to evaluate the percentiles of the χ^2 distribution with 1 degree of freedom.

22. Find ξ_p for an exponential random variable with parameter λ.

23. Express the percentiles of $Y = X^2$ in terms of the percentiles of X and the percentiles of $Z = \sqrt{X}$ in terms of those of X.

24. If X is uniform (0, 1), find the distribution function and density function for $Y = -2 \log X$ (log with base e).

25. If X is normal with mean μ and variance σ^2, show that

$$P(X \leq t) = N_Z\left(\frac{t - \mu}{\sigma}\right),$$

where N_Z is the standard normal distribution function ($\mu = 0$, $\sigma = 1$).

26. If X is a *positive random variable* ($F_X(0) = 0$), show that $E[X]E[1/X] \geq 1$.

1.4 Vector Random Variables

When treating several random variables simultaneously, all defined on the same sample space, it is frequently convenient to consider them as components of a vector and thus to consider random vectors. We shall use boldface type to denote vectors; thus $\mathbf{X}$ will denote a random vector of, say, k components $X_1, X_2, \ldots, X_k$, written as a column, and $\mathbf{x}$ will denote a column vector of observed values $x_1, x_2, \ldots, x_k$, for $\mathbf{X}$. The number of components $\mathbf{X}$ has will be clear from the context; each component X_i, of course, is a one-dimensional random variable considered alone. When discussing vector random variables in general, the first component X_1 could be discrete and the second component X_2 could be continuous. In the applications we shall consider, this sort of generality is not required and all of the components of $\mathbf{X}$ will either be discrete or all will be continuous. In the former case we shall say that $\mathbf{X}$ is discrete and, in the latter, that $\mathbf{X}$ is continuous.

If $\mathbf{X}$ is discrete, then it has a probability function, $p_{\mathbf{X}}$, which specifies the probabilities of occurrence of individual points $\mathbf{x}$ in a k-dimensional space (k is the number of components of $\mathbf{X}$). Its range $R_{\mathbf{X}}$ is the collection of k-tuples, $\mathbf{x}$, such that $p_{\mathbf{X}}(\mathbf{x}) > 0$. If $\mathbf{X}$ is continuous it has a density function, $f_{\mathbf{X}}$, defined on a k-dimensional space, which can be integrated over regions of the space to provide probabilities. Its range $R_{\mathbf{X}}$ is the collection of k-tuples $\mathbf{x}$ such that $f_{\mathbf{X}}(\mathbf{x}) > 0$. The marginal density or probability function for X_i, the ith component of $\mathbf{X}$, is derived by integrating $f_{\mathbf{X}}$, if $\mathbf{X}$ is continuous, or summing $p_{\mathbf{X}}$, if X is discrete, over the full ranges of all the other components of $\mathbf{X}$. The expected value of any function of $\mathbf{X}$, say $g(\mathbf{X})$, is given by

$$E[g(\mathbf{X})] = \sum_{\mathbf{x} \in R_{\mathbf{X}}} g(\mathbf{x}) p_{\mathbf{X}}(\mathbf{x}) \qquad \text{if } \mathbf{X} \text{ is discrete}$$

$$= \int_{R_{\mathbf{X}}} g(\mathbf{x}) f_{\mathbf{X}}(\mathbf{x})\, d\mathbf{x} \qquad \text{if } \mathbf{X} \text{ is continuous.}$$

Of special interest are the joint moments of the components of $\mathbf{X}$; the $(r_1, r_2, \ldots, r_k)$ moment of $\mathbf{X}$ is $E[X_1^{r_1} X_2^{r_2} \cdots X_k^{r_k}]$. Note that if $r_2 = r_3 = \cdots = r_k = 0$, then $E[X_1^{r_1} X_2^{r_2} \cdots X_k^{r_k}] = E[X_k^{r_1}]$ gives the (one-dimensional) moments of X_1; similarly, the joint moments include the (one-dimensional) moments of $X_2, X_3, \ldots, X_k$.

As in the case of one-dimensional random variables, the most frequently used moments are the first and second (mixed and pure). Define $E[\mathbf{X}]$ to be the $k \times 1$ vector having ith component $E[X_i]$. Then

$$E[\mathbf{X}] = \boldsymbol{\mu},$$

where $\boldsymbol{\mu}$ has as components the means of the components of $\mathbf{X}$. ($\boldsymbol{\mu}$ is called the *mean vector* of $\mathbf{X}$ or the vector of means for $\mathbf{X}$.) Also define $\mathbf{V} = E[(\mathbf{X} - \boldsymbol{\mu})(\mathbf{X} - \boldsymbol{\mu})']$ to be the $k \times k$ matrix with ij element $E[(X_i - \mu_i)(X_j - \mu_j)]$. $\mathbf{V}$ is called the *variance-covariance* matrix of $\mathbf{X}$ (frequently shortened to just *covariance* matrix), since its iith element is the variance of X_i and its ijth element is the covariance of X_i and X_j. We shall let σ_{ij} represent the ijth element of V, as a general notation and write $\mathbf{V} = \|\sigma_{ij}\|$ to express this. The *correlation* between X_i and X_j is

$$\rho_{ij} = \frac{\sigma_{ij}}{\sqrt{\sigma_{ii}}\sqrt{\sigma_{jj}}};$$

for any X_i and X_j, $|\rho_{ij}| \leq 1$.

The components of $\mathbf{X}$ are *independent* random variables if and only if

$$p_{\mathbf{X}}(\mathbf{x}) = \prod_{i=1}^{k} p_{X_i}(x_i), \qquad \text{for all } \mathbf{x}, \text{ if } \mathbf{X} \text{ is discrete}$$

$$f_{\mathbf{X}}(\mathbf{x}) = \prod_{i=1}^{k} f_{X_i}(x_i), \qquad \text{for all } \mathbf{x}, \text{ if } \mathbf{X} \text{ is continuous.}$$

Then, if the components of $\mathbf{X}$ are independent, the expected value of a multiplicative function of $X_1, X_2, \ldots, X_k$ is the product of the respective expected values; in particular, if the components of $\mathbf{X}$ are independent, the covariance of X_i and X_j is zero, for all $i \neq j$ and the covariance matrix is diagonal (which we express as $\text{diag}(\sigma_{11}, \sigma_{22}, \ldots, \sigma_{kk})$).

A *multinomial trial* is an experiment with k distinct possible outcomes. (Thus the roll of a single die is a multinomial trial with $k = 6$; the final grade which a student receives in a statistics course is a multinomial trial with $k = 5$, if the only possible grades are A, B, C, D, F.) In an experiment consisting of n repeated, independent, multinomial trials, where p_i, $i = 1, 2, \ldots, k$ is the probability of the occurrence of the ith outcome on a single trial, define X_i to be the number of times outcome i occurs, and let $\mathbf{X}$ be the column vector with components $X_1, X_2, \ldots, X_k$. Then $\mathbf{X}$ is the *multinomial random variable* (vector, actually); its range is $R_{\mathbf{X}} = \{(x_1, x_2, \ldots, x_k): x_i = 0, 1, \ldots, n, \ i = 1, 2, \ldots, k, \ \sum x_i = n\}$ and its probability function is

$$p_{\mathbf{X}}(\mathbf{x}) = \frac{n!}{\prod x_i!} p_1^{x_1} p_2^{x_2} \ldots p_k^{x_k}, \qquad \text{for } \mathbf{x} \in R_{\mathbf{X}}, \qquad 0 \leq p_i \leq 1, \sum p_i = 1;$$

its moment generating function is

$$m_{\mathbf{X}}(\mathbf{t}) = \left[\sum_{i=1}^{k} p_i e^{t_i}\right]^n;$$

the vector of means is given by $\boldsymbol{\mu}' = n(p_1, p_2, \ldots, p_k)$ and the covariance matrix $\mathbf{V}$ has diagonal elements $np_i(1 - p_i)$ and off diagonal elements $-np_ip_j$. The correlation, then, between X_i and X_j is

$$\rho_{ij} = \frac{-\sqrt{p_ip_j}}{\sqrt{(1 - p_i)(1 - p_j)}}.$$

For the multinomial random variable, as just defined, $\sum X_i = n$ and thus there is a linear dependency between the components of $\mathbf{X}$ (as is reflected by the requirement for $p_{\mathbf{X}}(\mathbf{x})$ that $\sum x_i = n$; this simply means that all of the probability in the k-dimensional space for $\mathbf{X}$ is actually lying on the face of the simplex defined by $x_i = 0, 1, 2, \ldots, n$ and $\sum x_i = n$ and all of the probability then lies in a $(k - 1)$-dimensional subspace). This is an example of a *degenerate random vector*, one with redundancy built into its coordinate system; the redundancy is used to simplify the notation for $p_{\mathbf{X}}$, $\boldsymbol{\mu}$, and $\mathbf{V}$. The degeneracy or singularity is also reflected in the fact that $\mathbf{V}$ is a singular matrix.

The *multivariate normal* random vector $\mathbf{X}$ has range

$$R_{\mathbf{X}} = \{(x_1, x_2, \ldots, x_k): \quad -\infty < x_i < \infty, i = 1, 2, \ldots, k\}$$

and density function

$$f_{\mathbf{X}}(\mathbf{x}) = \frac{|\mathbf{B}|^{1/2}}{(2\pi)^{k/2}} e^{-(1/2)(\mathbf{x}-\mathbf{a})'\mathbf{B}(\mathbf{x}-\mathbf{a})}, \qquad \mathbf{x} \in R_{\mathbf{X}}, \qquad \mathbf{a} \in R_{\mathbf{X}},$$

$\mathbf{B}$ is positive definite.

Since $\mathbf{B}$ is positive definite, there exists a nonsingular matrix $\mathbf{C}$ such that $\mathbf{C}'\mathbf{B}\mathbf{C} = \mathbf{I}$, the $k \times k$ identity matrix. Note as well then that $\mathbf{B} = (\mathbf{C}\mathbf{C}')^{-1}$, $|\mathbf{B}| = |\mathbf{C}|^{-2}$, and $|\mathbf{B}| \cdot |\mathbf{C}|^2 = 1$. To verify that $f_{\mathbf{X}}$ is a density function, in the integral

$$\int_{R_{\mathbf{X}}} \frac{|\mathbf{B}|^{1/2}}{(2\pi)^{k/2}} e^{-(1/2)(\mathbf{x}-\mathbf{a})'\mathbf{B}(\mathbf{x}-\mathbf{a})} \, d\mathbf{x}$$

transform to new variables of integration by $\mathbf{y} = \mathbf{C}^{-1}(\mathbf{x} - \mathbf{a})$. Note that $\mathbf{x} = \mathbf{C}\mathbf{y} + \mathbf{a}$. The jacobian of this transformation is $|\mathbf{C}|$ and the range of $\mathbf{y}$ is again $R_{\mathbf{X}}$, the whole k-dimensional space. Then, the integral becomes

$$\int_{R_{\mathbf{X}}} \frac{|\mathbf{B}|^{1/2}}{(2\pi)^{k/2}} e^{-(1/2)\mathbf{y}'\mathbf{C}'\mathbf{B}\mathbf{C}\mathbf{y}} |\mathbf{C}| \, d\mathbf{y} = \int_{R_{\mathbf{X}}} \frac{1}{(2\pi)^{k/2}} e^{-(1/2)\mathbf{y}'\mathbf{y}} \, d\mathbf{y}$$

$$= \prod_{i=1}^{k} \int_{-\infty}^{\infty} \frac{1}{\sqrt{2\pi}} e^{-y_i^2/2} \, dy_i = 1,$$

since each integral in the product has value 1; thus, since $f_{\mathbf{X}}$ is never negative and its integral is 1, it is in fact a density function. To derive the moment generating function for $\mathbf{X}$, we utilize the same transformation; thus

$$m_{\mathbf{X}}(\mathbf{t}) = E[e^{\mathbf{t}'\mathbf{x}}] = \int_{R_{\mathbf{X}}} e^{\mathbf{t}'\mathbf{x}} \frac{|\mathbf{B}|^{1/2}}{(2\pi)^{k/2}} e^{-(1/2)(\mathbf{x}-\mathbf{a})'\mathbf{B}(\mathbf{x}-\mathbf{a})} \, d\mathbf{x}$$

$$= \int_{R_{\mathbf{X}}} e^{\mathbf{t}'(\mathbf{C}\mathbf{y}+\mathbf{a})} \frac{|\mathbf{B}|^{1/2}}{(2\pi)^{k/2}} e^{-(1/2)\mathbf{y}'\mathbf{C}'\mathbf{B}\mathbf{C}\mathbf{y}} |\mathbf{C}| \, d\mathbf{y}$$

$$= e^{\mathbf{t}'\mathbf{a}} \int_{R_{\mathbf{X}}} \frac{1}{(2\pi)^{k/2}} e^{-(1/2)\{(\mathbf{y}'\mathbf{y}-2\mathbf{t}'\mathbf{C}\mathbf{y}\}} \, d\mathbf{y}.$$

Now,

$$\mathbf{y}'\mathbf{y} - 2\mathbf{t}'\mathbf{C}\mathbf{y} + \mathbf{t}'\mathbf{C}\mathbf{C}'\mathbf{t} - \mathbf{t}'\mathbf{C}\mathbf{C}'\mathbf{t} = (\mathbf{y} - \mathbf{C}'\mathbf{t})'(\mathbf{y} - \mathbf{C}'\mathbf{t}) - \mathbf{t}'\mathbf{C}\mathbf{C}'\mathbf{t};$$

Thus

$$m_{\mathbf{X}}(\mathbf{t}) = e^{\mathbf{t}'\mathbf{a}+(1/2)\mathbf{t}'\mathbf{C}\mathbf{C}'\mathbf{t}} \int_{R_{\mathbf{X}}} \frac{1}{(2\pi)^{k/2}} e^{-(1/2)(\mathbf{y}-\mathbf{C}'\mathbf{t})'(\mathbf{y}-\mathbf{C}'\mathbf{t})} \, d\mathbf{y}$$

$$= e^{\mathbf{t}'\mathbf{a}+(1/2)\mathbf{t}'\mathbf{C}\mathbf{C}'\mathbf{t}} = e^{\mathbf{t}'\mathbf{a}+(1/2)\mathbf{t}'\mathbf{B}^{-1}\mathbf{t}}.$$

The cumulant generating function for $\mathbf{X}$ is

$$c_{\mathbf{X}}(\mathbf{t}) = \log m_{\mathbf{X}}(\mathbf{t}) = \mathbf{t}'\mathbf{a} + (1/2)\mathbf{t}'\mathbf{B}^{-1}\mathbf{t};$$

realizing that the first derivatives of this function, with respect to the components of $\mathbf{t}$, evaluated at $\mathbf{t} = \mathbf{0}$, give the mean values of the components of

$\mathbf{X}$, it is easily seen that

$$\boldsymbol{\mu}_{\mathbf{X}} = \mathbf{a}.$$

Thus the vector $\mathbf{a}$ in $f_{\mathbf{X}}$ is the vector of means of $\mathbf{X}$; we shall henceforth use $\boldsymbol{\mu}$ to represent this vector. Similarly the second pure and mixed derivatives of $c_{\mathbf{X}}(\mathbf{t})$, evaluated at $\mathbf{t} = \mathbf{0}$, give the variances and covariances of the components of $\mathbf{X}$; these second derivatives, written in matrix form, simply give $\mathbf{B}^{-1}$. Thus $\mathbf{B}^{-1}$ is the variance-covariance matrix for $\mathbf{X}$; we shall use $\boldsymbol{\Sigma} = \|\sigma_{ij}\|$ to represent this matrix. Thus, the multivariate normal density will be written

$$f_{\mathbf{X}}(\mathbf{x}) = \frac{|\boldsymbol{\Sigma}|^{-1/2}}{(2\pi)^{k/2}} e^{-(1/2)(\mathbf{x}-\boldsymbol{\mu})'\boldsymbol{\Sigma}^{-1}(\mathbf{x}-\boldsymbol{\mu})};$$

the reader should note the great similarity between this formula and the univariate normal density:

$$\frac{1}{\sigma\sqrt{2\pi}} e^{-(1/2)(x-\mu)^2/\sigma^2} = \frac{(\sigma^2)^{-1/2}}{(2\pi)^{1/2}} e^{-(1/2)(x-\mu)(\sigma^2)^{-1}(x-\mu)}.$$

By $\mathbf{X} \sim MVN(\boldsymbol{\mu}, \boldsymbol{\Sigma})$, we shall mean $\mathbf{X}$ is a multivariate normal vector with mean $\boldsymbol{\mu}$, covariance $\boldsymbol{\Sigma}$.

The marginal moment generating function for X_1 is

$$\begin{aligned} m_{X_1}(t) &= E(e^{tX_1+0X_2+\cdots+0X_k}) \\ &= E(e^{\mathbf{t}'_1\mathbf{X}}) \\ &= m_{\mathbf{X}}(\mathbf{t}_1), \end{aligned}$$

where $\mathbf{t}'_1 = (t, 0, 0, \ldots, 0)$. But

$$\begin{aligned} m_{\mathbf{X}}(\mathbf{t}_1) &= e^{\mathbf{t}'_1\boldsymbol{\mu}_{\mathbf{X}}+(1/2)\mathbf{t}_1'\boldsymbol{\Sigma}\mathbf{t}_1} \\ &= e^{t\mu_1+(1/2)t^2\sigma_{11}}; \end{aligned}$$

thus X_1 is univariate normal with mean μ_1 and variance σ_{11}, since its moment generating function is of the normal form. Similarly, X_i is univariate normal with mean μ_i and variance σ_{ii}, $i = 2, 3, \ldots, k$.

In fact, the moment generating function for $\mathbf{Y} = \mathbf{DX}$, where $\mathbf{D}$ is a $p \times k$ matrix of rank p, is

$$\begin{aligned} m_{\mathbf{Y}}(\mathbf{t}) &= E(e^{\mathbf{t}'\mathbf{DX}}) \\ &= m_{\mathbf{X}}(\mathbf{D}'\mathbf{t}) \\ &= e^{\mathbf{t}'\mathbf{D}\boldsymbol{\mu}_{\mathbf{X}}+(1/2)\mathbf{t}'\mathbf{D}\boldsymbol{\Sigma}\mathbf{D}'\mathbf{t}} \end{aligned}$$

and thus $\mathbf{Y}$ is multivariate normal with mean $\mathbf{D}\boldsymbol{\mu}_{\mathbf{X}}$ and covariance matrix $\mathbf{D}\boldsymbol{\Sigma}\mathbf{D}'$, since its moment generating function is of that form.

If $\boldsymbol{\Sigma} = \text{diag}(\sigma_{11}, \sigma_{22}, \ldots, \sigma_{kk})$, note that

$$(\mathbf{x} - \boldsymbol{\mu}_{\mathbf{X}})'\boldsymbol{\Sigma}^{-1}(\mathbf{x} - \boldsymbol{\mu}_{\mathbf{X}}) = \sum_{i=1}^{k} \frac{(x_i - \mu_i)^2}{\sigma_{ii}}$$

and that $|\boldsymbol{\Sigma}|^{-1/2} = \left[\prod_{i=1}^{k} \sigma_{ii}\right]^{-1/2}$; thus, we would have

$$f_{\mathbf{X}}(\mathbf{x}) = \frac{\left\{\prod_{i=1}^{k} \sigma_{ii}\right\}^{-1/2}}{(2\pi)^{k/2}} e^{-\sum_{i=1}^{k} (x_i-\mu_i)^2/2\sigma_{ii}}$$

$$= \prod_{i=1}^{k} \frac{1}{(2\pi\sigma_{ii})^{1/2}} e^{-(x_i-\mu_i)^2/2\sigma_{ii}} = \prod_{i=1}^{k} f_{X_i}(x_i)$$

and the components of $\mathbf{X}$ are independent random variables. Conversely, if the components of $\mathbf{X}$ are independent, then the covariances are all zero and $\boldsymbol{\Sigma}$ is diagonal. Thus if $\mathbf{X}$ is multivariate normal the individual components of $\mathbf{X}$ are independent scalar random variables if and only if $\boldsymbol{\Sigma}$ is a diagonal matrix.

Since $\boldsymbol{\Sigma}$ is a positive definite matrix, there exists a nonsingular matrix $\mathbf{C}$ (not unique) such that $\mathbf{C}\boldsymbol{\Sigma}\mathbf{C}' = \mathbf{I}$; again then $\boldsymbol{\Sigma} = (\mathbf{C}'\mathbf{C})^{-1}$ and if $\mathbf{Y} = \mathbf{C}\mathbf{X}$ from above we see that $\mathbf{Y}$ is multivariate normal with mean $\mathbf{C}\boldsymbol{\mu}$ and covariance matrix $\mathbf{C}\boldsymbol{\Sigma}\mathbf{C}' = \mathbf{I}$. Thus the components of $\mathbf{Y}$ are independent, each with variance 1. This type of transformation proves useful in deriving many important results. Notice also that if $\mathbf{C}$ is partitioned into $\begin{pmatrix}\mathbf{C}_1\\ \mathbf{C}_2\end{pmatrix}$ where $\mathbf{C}_1$ is $q \times k$, and thus $\mathbf{Y} = \begin{pmatrix}\mathbf{Y}_1\\ \mathbf{Y}_2\end{pmatrix} = \begin{pmatrix}\mathbf{C}_1\mathbf{X}\\ \mathbf{C}_2\mathbf{X}\end{pmatrix}$, then $\mathbf{Y}_1$ is $q \times 1$ multivariate normal with mean vector $\mathbf{C}_1\boldsymbol{\mu}$, covariance $\mathbf{I}$.

Given a discrete random vector $\mathbf{X}$ with k components, let $\mathbf{X}_1$ be the subvector of the first q components of $\mathbf{X}$ and let $\mathbf{X}_2$ be the remaining $k - q$ components of $\mathbf{X}$. The conditional probability that $\mathbf{X}_1 = \mathbf{x}_1$, given that $\mathbf{X}_2 = \mathbf{x}_2$, is

$$p_{\mathbf{X}_1|\mathbf{X}_2}(\mathbf{x}_1 \mid \mathbf{x}_2) = \frac{p_{\mathbf{X}}(\mathbf{x})}{p_{\mathbf{X}_2}(\mathbf{x}_2)}$$

where $\mathbf{x} = \begin{pmatrix}\mathbf{x}_1\\ \mathbf{x}_2\end{pmatrix} \in R_{\mathbf{X}}$, and $p_{\mathbf{X}_2}(\mathbf{x}_2)$ is the marginal probability function for $\mathbf{X}_2$ evaluated at the given values $\mathbf{x}_2$. This definition follows directly from the definition of conditional probability given in section 1.2, letting A be the event that $\mathbf{X}_1 = \mathbf{x}_1$ and B the event that $\mathbf{X}_2 = \mathbf{x}_2$; $A \cap B$ then is the event that $\mathbf{X}_1 = \mathbf{x}_1$ and $\mathbf{X}_2 = \mathbf{x}_2$, i.e., $\mathbf{X} = \mathbf{x}$ where $\mathbf{x} = \begin{pmatrix}\mathbf{x}_1\\ \mathbf{x}_2\end{pmatrix}$. Let $\mathbf{X}$ now be a continuous random vector of k components and, again, partition $\mathbf{X}$ into $\mathbf{X}_1$, its first q components, and $\mathbf{X}_2$, its last $k - q$ components. Then the conditional density for $\mathbf{X}_1$, given $\mathbf{X}_2 = \mathbf{x}_2$, is

$$f_{\mathbf{X}_1|\mathbf{X}_2}(\mathbf{x}_1 \mid \mathbf{x}_2) = \frac{f_{\mathbf{X}}(\mathbf{x})}{f_{\mathbf{X}_2}(\mathbf{x}_2)}$$

where $\mathbf{x} = \begin{pmatrix} \mathbf{x}_1 \\ \mathbf{x}_2 \end{pmatrix} \in R_{\mathbf{X}}$ and $f_{\mathbf{X}_2}(\mathbf{x}_2)$ is the marginal density for $\mathbf{X}_2$ evaluated at the given values $\mathbf{x}_2$. Note that this is exactly the same definition given earlier for $\mathbf{X}$ discrete, with probability functions replaced by density functions. One way of seeing why the definition should take this form in the continuous case is to realize that $f_{\mathbf{X}}(\mathbf{x})$ defines a k-dimensional surface in a $(k + 1)$-dimensional space. If we are given that $\mathbf{X}_2 = \mathbf{x}_2$, then we are certain that some outcome on the $k - q$ dimensional hyperplane $\mathbf{X}_2 = \mathbf{x}_2$ occurred when the experiment was performed. To compute probabilities that $\mathbf{X}_1$ lies in specific regions in its q-dimensional space, it would seem natural to take the intersection of the surface $f_{\mathbf{X}}(\mathbf{x})$ with the hyperplane $\mathbf{X}_2 = \mathbf{x}_2$ and to use the resulting q-dimensional surface to compute the probability that $\mathbf{X}_1$ lies in any given region. The above definition accomplishes exactly this, where the division by $f_{\mathbf{X}_2}(\mathbf{x}_2)$ is done to assure that we are in fact using a legitimate density function.

Example 1.4.1

Assume that a weekly customer on a sport fishing boat catches 0, 1, 2, or 3 fish every time he goes out, and that, each time, the probabilities of these numbers occurring are 1/2, 1/4, 1/8, and 1/8, respectively. Thus, each time he goes fishing on this boat can be thought of as a multinomial trial with $k = 4$ possible outcomes. Assume that in a two-month period he goes on this boat $n = 9$ times and that the results from week to week are independent. Then, if $\mathbf{X}' = (X_1, X_2, X_3, X_4)$ is the vector of the number of times, out of the 9, that he caught 0, 1, 2, and 3 fish, respectively, $\mathbf{X}$ is multinomial with $n = 9$, $p_1 = 1/2$, $p_2 = 1/4$, $p_3 = p_4 = 1/8$, and

$$p_{\mathbf{X}}(\mathbf{x}) = \frac{9!}{x_1!x_2!x_3!x_4!}\left(\frac{1}{2}\right)^{x_1}\left(\frac{1}{4}\right)^{x_2}\left(\frac{1}{8}\right)^{x_3}\left(\frac{1}{8}\right)^{x_4}.$$

The conditional probability function of $\mathbf{X}_2' = (X_2, X_3, X_4)$, given $X_1 = 2$, is

$$p_{\mathbf{X}_2|X_1}(\mathbf{x}_2 \mid x_1) = \frac{\dfrac{9!}{2!x_2!x_3!x_4!}\left(\dfrac{1}{2}\right)^{2}\left(\dfrac{1}{4}\right)^{x_2}\left(\dfrac{1}{8}\right)^{x_3}\left(\dfrac{1}{8}\right)^{x_4}}{\dbinom{9}{2}\left(\dfrac{1}{2}\right)^{2}\left(1-\dfrac{1}{2}\right)^{7}}$$

$$= \frac{7!2^7}{x_2!x_3!x_4!}\left(\frac{1}{4}\right)^{x_2}\left(\frac{1}{8}\right)^{x_3}\left(\frac{1}{8}\right)^{x_4},$$

since the marginal probability function for X_1 is binomial with $n = 9$, $p = 1/2$ (see problem 3 below). Note that the conditional probability function for $\mathbf{X}_2$, given $X_1 = x_1$, is again multinomial in form, which is the case in general.

Example 1.4.2

Assume that $\mathbf{X}$ is 2×1, bivariate normal with $\boldsymbol{\mu} = \begin{pmatrix} \mu_1 \\ \mu_2 \end{pmatrix}$, $\boldsymbol{\Sigma} = \begin{pmatrix} \sigma_{11} & \sigma_{12} \\ \sigma_{12} & \sigma_{22} \end{pmatrix}$. Then

$$|\boldsymbol{\Sigma}| = \sigma_{11}\sigma_{22} - \sigma_{12}^2 = \sigma_{11}\sigma_{22}\left(1 - \frac{\sigma_{12}^2}{\sigma_{11}\sigma_{22}}\right) = \sigma_{11}\sigma_{22}(1 - \rho^2),$$

$$\boldsymbol{\Sigma}^{-1} = \frac{1}{\sigma_{11}\sigma_{22}(1 - \rho^2)}\begin{pmatrix} \sigma_{22} & -\sigma_{12} \\ -\sigma_{12} & \sigma_{11} \end{pmatrix}$$

$$= \frac{1}{1 - \rho^2}\begin{pmatrix} \dfrac{1}{\sigma_{11}} & -\dfrac{\rho}{\sqrt{\sigma_{11}\sigma_{22}}} \\ -\dfrac{\rho}{\sqrt{\sigma_{11}\sigma_{22}}} & \dfrac{1}{\sigma_{22}} \end{pmatrix}$$

and the joint density of X_1 and X_2 is

$$f_{\mathbf{X}}(\mathbf{x}) = \frac{|\boldsymbol{\Sigma}|^{-1/2}}{2\pi} e^{-(1/2)(\mathbf{x}-\boldsymbol{\mu})'\Sigma^{-1}(\mathbf{x}-\boldsymbol{\mu})}$$

$$= \frac{1}{2\pi\sqrt{\sigma_{11}\sigma_{22}(1 - \rho^2)}} e^{-\frac{1}{2(1-\rho^2)}\left\{\frac{(x_1-\mu_1)^2}{\sigma_{11}} + \frac{(x_2-\mu_2)^2}{\sigma_{22}} - 2\rho\frac{(x_1-\mu_1)}{\sqrt{\sigma_{11}}}\frac{(x_2-\mu_2)}{\sqrt{\sigma_{22}}}\right\}}.$$

Then, if we are given that $X_2 = x_2$, some particular value, the conditional density of X_1 is

$$f_{X_1|X_2}(x_1 \mid x_2) = \frac{f_{\mathbf{X}}(\mathbf{x})}{f_{X_2}(x_2)}$$

$$= \frac{1}{\sqrt{2\pi\sigma_{11}(1 - \rho^2)}} e^{-1/2\frac{\{x_1-\mu_1-\rho\sqrt{(\sigma_{11}/\sigma_{22})}(x_2-\mu_2)\}^2}{\sigma_{11}(1-\rho^2)}},$$

and thus the conditional distribution of X_1 given $X_2 = x_2$ is again normal with mean $\mu_1 + \rho\sqrt{\sigma_{11}/\sigma_{22}}(x_2 - \mu_2)$ and variance $\sigma_{11}(1 - \rho^2)$. Notice then that the conditional mean of X_1 depends upon the given value of X_2, if $\rho \neq 0$, but that the conditional variance of X_1 does not. Whether we are given $X_2 = 0$ or $X_2 = 100$, the conditional variance of X_1 is unchanged. The conditional mean value of X_1, given $X_2 = x_2$, is called the *regression* of X_1 on X_2.

Two important results from probability theory, that will be of use to us in our study of mathematical statistics, are the *law of large numbers* and the *central limit theorem.* The *law of large numbers* states that if $X_1, X_2, \ldots, X_k, \ldots$ is a sequence of independent, identically distributed random variables, each with mean μ and variance σ^2, then

$$\lim_{n\to\infty} P(|\bar{X}_n - \mu| > \varepsilon) = 0, \qquad \text{for any } \varepsilon > 0, \qquad \text{where } \bar{X}_n = \frac{1}{n}\sum_{i=1}^{n} X_i.$$

It is important to realize that the law of large numbers makes a statement about the limit of a sequence of probabilities. There is no guarantee at all that $\bar{X}_n = \mu$ for any n, nor that they are equal in value in the limit. However, the probability that the distance between $\bar{X}_n$ and μ exceeds ε, no matter how small, converges to 0 as n gets larger; thus, it is likely that the value of $\bar{X}_n$, for large n, is close to μ.

The central limit theorem, on the other hand, states that

$$(\bar{X}_n - \mu)/(\sigma/\sqrt{n}) = (\sum X_i - n\mu)/\sigma\sqrt{n}$$

has the standard normal distribution in the limit as n gets large. Thus, if we observe a random variable X, whose value is the sum of a large number of independent random variables, we have some reason to believe that X is normally distributed.

EXERCISE 1.4

1. If a fair die is rolled 6 times (or each of 6 fair dice is rolled once) what is the probability that all 6 faces occur? that the same face occurs 6 times?

2. In question 1, what is the probability that face 2 occurs (exactly) twice as often as face 1?

3. Show that, if $\mathbf{X}$ is multinomial with parameters $n, p_1, p_2, \ldots, p_k$, X_i is binomial with parameters n and p_i.

4. Show that if $\mathbf{X}$ is $k \times 1$, multivariate normal with mean $\boldsymbol{\mu}$ and covariance matrix $\boldsymbol{\Sigma}$, then any linear combination $Y = \mathbf{a}'\mathbf{X}$, where $\mathbf{a}' = (a_1, a_2, \ldots, a_k)$ is a vector of constants, is a scalar normal random variable with mean $\mathbf{a}'\boldsymbol{\mu}$ and variance $\mathbf{a}'\boldsymbol{\Sigma}\mathbf{a}$. This is called the *reproductive property* of the normal distribution.

5. An important special case of problem 4 is given by $\boldsymbol{\mu}' = (\mu, \mu, \ldots, \mu)$, $\boldsymbol{\Sigma} = \sigma^2\mathbf{I}$ (the identity matrix), with $\mathbf{a}' = (1/n, 1/n, \ldots, 1/n)$. Thus, the components of $\mathbf{X}$ are independent, each with the same mean μ and same variance σ^2, and $Y = \mathbf{a}'\mathbf{X} = \bar{X} = (1/n)\sum_{i=1}^{n} X_i$ is the arithmetic average or mean of the components of $\mathbf{X}$. What are the mean and variance of $\bar{X}$?

6. As in problem 5 (equal means, variances, independence) let $Z = \mathbf{b}'\mathbf{X}$, where $\mathbf{b}$ is any $n \times 1$ vector of constants. If we require $E[Z] = \mu$, then $\sum_{i=1}^{n} b_i = 1$; show that among all choices of **b** such that $\sum_{i=1}^{n} b_i = 1$, the variance of Z is minimized with $b_i = 1/n$ for all i.

7. Show that the covariance matrix $\mathbf{V}$ of the multinomial random variable $\mathbf{X}$ is singular. (Hint: look at the sum of the elements across each row.)

8. Show that if $X_1, X_2, \ldots, X_n$ are independent random variables, then

$$m_{\sum_{i=1}^{n} X_i}(t) = \prod_{i=1}^{n} m_{X_i}(t).$$

9. If $X_1, X_2, \ldots, X_n$ are independent, Poisson random variables with parameters $\lambda_1, \lambda_2, \ldots, \lambda_n$, respectively, show that $Y = \sum_{i=1}^{n} X_i$ is again Poisson with parameter $\nu = \sum \lambda_i$. Thus the Poisson random variable also reproduces itself.

10. Let $X_1, X_2, \ldots, X_k$ be independent, binomial random variables, where $E[X_i] = n_i p$, $i = 1, 2, \ldots, k$. Show that $Y = \sum_{i=1}^{k} X_i$ is again binomial with parameters $n = \sum_{i=1}^{k} n_i$ and p, and in a restricted sense, the binomial reproduces itself.

11. Let $X_1, X_2, \ldots, X_r$ be independent, exponential random variables, each with the same parameter λ. Show that $Y = \sum_{i=1}^{r} X_i$ is a gamma random variable with parameters r and λ. Explain, in the context of a Poisson process, why this must be so.

12. Let X_1 and X_2 be independent, uniform (0, 1) random variables and derive the density function for $Y = X_1 + X_2$.

13. Let $X_1, X_2, \ldots, X_n$ be independent, gamma random variables, such that $E[X_i] = r_i/\lambda$, $i = 1, 2, \ldots, n$. Show that $Y = \sum_{i=1}^{n} X_i$ is a gamma random variable with parameters $r = \sum_{i=1}^{n} r_i$, λ. Explain the significance of this result in terms of a Poisson process.

14. Assume that $\mathbf{X}_1, \mathbf{X}_2, \ldots, \mathbf{X}_m$ are independent, multinomial random variables, with parameters $n_i, p_1, p_2, \ldots, p_k$, $i = 1, 2, \ldots, m$, respectively. Show that $\mathbf{Y} = \sum_{i=1}^{m} \mathbf{X}_i$ is multinomial with parameters $n = \sum_{i=1}^{m} n_i, p_1, p_2, \ldots, p_k$.

15. Assume that $\mathbf{X}_1, \mathbf{X}_2, \ldots, \mathbf{X}_n$ are independent, $p \times 1$, multivariate normal vectors with mean vector $\boldsymbol{\mu}_i$ and covariance matrix $\boldsymbol{\Sigma}_i$, $i = 1, 2, \ldots, n$, respectively. Show that $\mathbf{Y} = \sum_{i=1}^{n} a_i \mathbf{X}_i$ is again multivariate normal with mean vector $\sum a_i \boldsymbol{\mu}_i$ and covariance matrix $\sum_{i=1}^{n} a_i^2 \boldsymbol{\Sigma}_i$, where $a_1, a_2, \ldots, a_n$ is any set of constants. Hence, the multivariate normal also reproduces itself.

16. Let $\mathbf{X}$ be a $p \times 1$ multivariate normal vector with $\boldsymbol{\mu}$ and $\boldsymbol{\Sigma}$. Partition $\mathbf{X}$ into $\mathbf{X}_1$ and $\mathbf{X}_2$, where $\mathbf{X}_1$ consists of the first $q < p$ components of $\mathbf{X}$. Also partition $\boldsymbol{\mu}$ and $\boldsymbol{\Sigma}$:

$$\boldsymbol{\mu} = \begin{pmatrix} \boldsymbol{\mu}_1 \\ \boldsymbol{\mu}_2 \end{pmatrix}, \qquad \boldsymbol{\Sigma} = \begin{pmatrix} \boldsymbol{\Sigma}_{11} & \boldsymbol{\Sigma}_{12} \\ \boldsymbol{\Sigma}_{21} & \boldsymbol{\Sigma}_{22} \end{pmatrix}$$

where $\boldsymbol{\mu}_1$ is $q \times 1$ and $\boldsymbol{\Sigma}_{11}$ is $q \times q$. Show that the density of $\mathbf{X}_1$, given $\mathbf{X}_2 = \mathbf{x}_2$, is multivariate normal with mean vector $\boldsymbol{\mu}_1 + \boldsymbol{\Sigma}_{12}\boldsymbol{\Sigma}_{22}^{-1}(\mathbf{x}_2 - \boldsymbol{\mu}_2)$ and covariance matrix $\boldsymbol{\Sigma}_{11} - \boldsymbol{\Sigma}_{12}\boldsymbol{\Sigma}_{22}^{-1}\boldsymbol{\Sigma}_{21}$. (Hint: The quadratic form in $f_{\mathbf{X}}$ is $(\mathbf{x} - \boldsymbol{\mu})'\boldsymbol{\Sigma}^{-1}(\mathbf{x} - \boldsymbol{\mu}) =$ $(\mathbf{x} - \boldsymbol{\mu})'\mathbf{D}'(\mathbf{D}\boldsymbol{\Sigma}\mathbf{D}')^{-1}\mathbf{D}(\mathbf{x} - \boldsymbol{\mu})$ where $\mathbf{D} = \begin{pmatrix} \mathbf{I} & -\boldsymbol{\Sigma}_{12}\boldsymbol{\Sigma}_{22}^{-1} \\ \mathbf{0} & \mathbf{I} \end{pmatrix}$; furthermore, since $|\mathbf{D}| = 1$, $|\boldsymbol{\Sigma}|^{-1/2} = |\mathbf{D}\boldsymbol{\Sigma}\mathbf{D}'|^{-1/2}$. This gives a handy form for writing $f_{\mathbf{X}}$.)

17. A $p \times 1$ *singular multivariate normal* vector $\mathbf{X}$ is one whose covariance matrix is singular. As in the case of a multinomial random variable, this means that there is redundancy in $\mathbf{X}$ and that one or more of the components of $\mathbf{X}$ can be written as a linear combination of the rest; thus all the probability for $\mathbf{X}$ is concentrated in a lower dimension subspace and, since $\boldsymbol{\Sigma}^{-1}$ does not exist, the density of $\mathbf{X}$ in p space does not exist. Show that if $\mathbf{Y}$ is $m \times 1$, multivariate normal with $\boldsymbol{\mu}$, $\mathbf{T}$ (non-singular) and $\mathbf{B}$ is a $p \times m$, $p > m$, matrix of rank m, $\mathbf{a}$ is a $p \times 1$ vector of constants, then $\mathbf{X} = \mathbf{BY} + \mathbf{a}$ has a singular normal distribution and find the mean vector for $\mathbf{X}$ and the variance covariance matrix for $\mathbf{X}$ in terms of $\boldsymbol{\mu}$ and $\boldsymbol{\Sigma}$.

18. Assume X is exponential with parameter λ and let a and b be positive constants. Compute $P(X > a + b \mid X > b)$ and compare this with $P(X > a)$. The exponential random variable is said to have no memory, because of this relationship; it can be shown to be the only continuous random variable for which this holds true.

19. Let X be a geometric random variable and a and b are positive integers. Compute $P(X > a + b \mid X > b)$ and compare with $P(X > a)$. The geometric random variable is the only discrete random variable with no memory.

20. Assume that a photographic plate is exposed to a radioactive source for an interval of time of length t. Given that n particles were emitted during this period of time, the number to strike the plate is binomial with parameters n and p. We assume that Y, the number emitted during this time is Poisson with parameter λt. Thus $P(X = k \mid Y = n) = \binom{n}{k} p^k q^{n-k}$, $k = 0, 1, 2, \ldots, n$. Find the unconditional probability function for X.

21. Assume that $X_1, X_2, \ldots, X_n$ are independent, Bernoulli random variables, each with parameter p. What is the distribution of $Y = \sum_{i=1}^{n} X_i$?

22. Assume that $X_1, X_2, \ldots, X_n$ are independent, random variables, each with the same distribution function $F_X(t)$. Define

$$X_{(n)} = \max(X_1, X_2, \ldots, X_n)$$

$$X_{(1)} = \min(X_1, X_2, \ldots, X_n);$$

thus $X_{(1)}$ is the smallest sample value and $X_{(n)}$ is the largest. Express $F_{X_{(1)}}(t)$ and $F_{X_{(n)}}(t)$ in terms of $F_X(t)$.

23. In problem 22 assume the X_i's are continuous and derive the density functions for $X_{(1)}$ and $X_{(n)}$.

24. Assume, in problem 22, that the X_i's are each exponential with parameter λ. Show that $X_{(1)}$ is exponential with parameter $n\lambda$.

25. Assume, in problem 22, that $X_1, X_2, \ldots, X_n$ are each geometric with parameter p. Show that $X_{(1)}$ is geometric with parameter $1 - q^n$.

26. Given $\mathbf{\Sigma} = \begin{pmatrix} \sigma_{11} & \sigma_{12} \\ \sigma_{12} & \sigma_{22} \end{pmatrix}$ is positive definite, find a non-singular matrix **C** such that $\mathbf{C\Sigma C} = \mathbf{I}$. Show how any number of other matrices with this same property may be defined. (Hint: Look at the eigenvectors of $\mathbf{\Sigma}$.)

27. Generalize problem 26 to a $k \times k$ matrix.

CHAPTER 2

Populations and Samples

2.1 Introduction

In this chapter we shall study some of the basic notions that are commonly used in applying probability theory to practical problems of inference. It is unfortunate, but perhaps necessary, that the application of statistics is still something of an art; the choice of the model to use and, thus, the appropriate statistical method to employ, is not always obvious. Indeed there are problems in which knowledgeable people cannot agree upon the "correct" model and method. Thus, it is not surprising perhaps that many students find statistics a confusing jumble of possibly conflicting ways of deriving an answer to a given practical problem.

Generally, it is much easier to compare two or more methods of answering a given question if the models used by the methods are clearly understood; the set of assumptions about the random variables occurring in the problem and their interrelationships, as well as the way in which they are assumed to represent practical reality, is called the model of the situation. Clearly, if two different models are assumed for the same problem it would not be surprising if they called for the use of different statistical methods and, conceivably, then led to different conclusions for the practical problem. Choosing the "correct" statistical method, thus, really comes down to the choice of the "correct" model of reality; as long as people are still free to disagree about the correct method to improve society, they are still free to disagree about the "correct" model for certain aspects of reality. Thus, there will probably always be disagreement, and room for imagination and art, in the application of statistics.

In this chapter we shall study the basic idea of a population (or a population distribution) and the concept of sampling from that population.

The inherent idea, of course, is to use known facts from the sample to make inferences about the population.

2.2 Populations and Samples

The word population is in common usage in everyday language. We are all familiar with such things as the population of the United States, the population of black bears on the North American continent, and the population of fish in Lake Superior. For any given point in time, each of these seems to be a well-defined entity, a set of individuals each of whom could be identified. Of course, as time passes, each of these populations changes, conceivably, through birth, death, and mobility. It is important to realize, then, in the contexts of such populations, that the character of the population may well vary with time, but for any given instant, such a population is in theory well defined.

Frequently, in applied problems, a specific aspect of the individuals in the population is the quantity to be studied. For example, we might want to estimate the average income of the population of the United States in 1970; then, rather than thinking of the population of the United States, as individual people, it is useful to consider only the incomes in 1970 of these individuals and, therefore, to consider the population of incomes in 1970; this, then, is a population of numbers, the numbers being the respective incomes of the individuals in the population. It might be of interest to a zoologist to estimate the average age, at a given time, of all the black bears on the North American continent or the average weight of all the fish in Lake Superior. Again, in these two examples, then, it is of interest to consider populations of numbers (ages and weights), rather than the populations of individual bears or fish. Frequently, rather than a single number (scalar) associated with each individual in a population, we may be interested in a vector of several numerical characteristics; thus, we might be interested in the yearly income, age, educational level, etc., of the individuals in the United States, or the age, height, and weight of the black bears in North America, or the weight, fin length, and girth of the fish in Lake Superior. Thus, it is not necessary that we be concerned only with populations of scalars; indeed, in more and more problems it is necessary to consider populations of vectors.

In each of the three examples just discussed, the population of numbers of interest was derived from a concrete population of individuals which definitely exist in a real sense. Many practical problems are concerned with various aspects of such populations. However, if we were to restrict our attention to only this type of population we would be ignoring a very large body of important problems concerned with *conceptual populations*, ones which the mind can conceive of but which do not exist in the same concrete manner as those discussed above. For example, consider a specific process for manufacturing 60-watt light bulbs. Conceivably, any number of such bulbs could be produced in exactly the same way and each such light bulb would have some length of time that it would burn, without failure, in a given type of usage (in a home, or a factory, or on board a boat, etc.). We then can conceive of the population of lengths of time that this type of bulb would last in this usage, a conceptual population, and can pose problems that are of interest regarding this conceptual population. As a second example, consider a specific usage of a given drug in the treatment of a given ailment in a human being, say tetracycline and its effect on a given type of skin infection. It is possible to conceive of any number of individuals, each with the same skin infection, being treated in the same way with the same dosage of tetracycline. Each person receiving the treatment might be looked at as being cured (0), controlled from further spread (1), or not affected by the drug (2). Thus we could conceive of a population of numbers, each being 0 or 1 or 2, generated from such treatments; indeed, for any given dosage level of tetracycline we can conceive of such a population and would probably be quite interested in how the proportions of 0's, 1's, and 2's were affected by the changes in the dosage.

Thus, our usage of the word *population* will be quite general. We may have in mind either a concrete population that really exists or a conceptual population which does not and never will exist. In either case we will almost exclusively be concerned with populations of scalars or vectors. Most statistical methods can be viewed as ways of making inferences about a population. Such inferences are made after the examination of a *sample* from the population. For example, we might be concerned with making a guess about the overall average value of all the numbers in the population (point estimation), or in giving an interval of values with a known probability of including the overall average (confidence interval), or in deciding whether a specific number is equal to the overall average (testing a hypothesis). In each of these cases we desire an answer based on only partial information, a sample from the population, rather than from a complete examination of the population itself.

A population can be either discrete or continuous, depending upon whether the set of numbers referred to is discrete or continuous. Continuous populations can occur in two ways. If we discuss the population of weights of people in the United States at a given time, then, since there is a finite number of people living there, the population of weights is truly discrete. However there is such a large number of distinct values that can occur as the weights of the individuals concerned that it is convenient to idealize the population and to think of the population of weights as being continuous;

in that way continuous techniques can be applied which are considerably simpler to use than are their discrete counterparts. Conceptual populations of time, distance, etc., shall also be considered continuous; if we really take a sample from such a conceptual population, the values which we can observe are again truly necessarily discrete, because of the discreteness of the measuring instrument used. Again, however, the greater simplicity of continuous techniques makes it advantageous to consider the population, and sample values, as being continuous.

For any given population of numbers, there is a discrete probability function or continuous density function which describes the distribution of values in the population, depending upon whether the population is discrete or continuous. The random variable X which has the given probability or density function is called the *population random variable*. Population random variables can be either discrete or continuous; the range of the population random variable thus is the set of values in the population.

Example 2.2.1

Consider the population generated by the treatment of skin infections using a given dosage of tetracycline, discussed above; if p_1 is the proportion of 0's, p_2 the proportion of 1's and p_3 the proportion of 2's, then the population random variable X is discrete with probability function

$$p_X(0) = p_1, \qquad p_X(1) = p_2, \qquad p_X(2) = p_3.$$

Example 2.2.2

Assume that the distribution of weights of people in the United States is well approximated by a normal density with $\mu = 150$, $\sigma = 30$. (By this we mean the proportion of people having weights in the interval (a, b) is essentially the same as the area under the given normal density, for all possible (a, b).) Then the population random variable X is normal with $\mu_X = 150$, $\sigma_X = 30$. Note as well that the true range of the population is finite; surely the interval from 0 to 1000 pounds, say, should encompass all the weights in the population; yet the normal density we have assumed is actually positive over the whole real line. The use of the normal density in such a case represents a further idealization, which, for all practical purposes, should not seriously affect any desired manipulations. Note that if X is normal with $\mu_X = 150$, $\sigma_X = 30$, then $f_X(0)$ and $f_X(1000)$ are zero to many decimal places, although still strictly positive.

There are many ways in which a sample can be selected from a population. It could be selected quite purposefully; if we are going to select a sample of, say, 10 values from the population, we could quite deliberately attempt to find the 10 smallest values that exist in the population. Clearly, such a purposeful way of selecting the sample would not in general be of much use in making inferences about the population as a whole, although it would give us information about the smallest population value. Granted that we do in general want to make inferences about the population as a whole, we should find other methods of selecting the sample.

As a general rule, samples that are selected haphazardly or randomly from the population have great advantages for making inferences about the population. Let us now discuss an important definition.

DEFINITION 2.2.1. A *random sample of size n of a random variable X* is a collection of n random variables, $X_1, X_2, \ldots, X_n$ such that

(i) $X_1, X_2, \ldots, X_n$ are independent.
(ii) $F_{X_i}(t) = F_X(t)$, for all t, $i = 1, 2, \ldots, n$.

In sampling at random from any finite population without replacement, we do not satisfy Definition 2.2.1. Suppose we have a finite population consisting of the numbers 1 through N and are going to select a sample of $n = 2$ of these values; thus, if X is the population random variable,

$$p_X(x) = \frac{1}{N}, \qquad x = 1, 2, \ldots, N.$$

There are $\binom{N}{2}$ different samples of size 2 that could be selected, and, if we choose any of the numbers in the population at random as our first sample value, and then choose at random any of the remaining $N - 1$ numbers as our second sample value, clearly each of the $\binom{N}{2}$ different possible pairs of values has the same probability of occurrence, and this would be called a *random sample* of size 2 *from the population.* However, if we let X_1 be the first value selected and X_2 the second, clearly the conditional probability function for X_2 depends upon the particular value X_1 takes on; thus X_1 and X_2 are not independent, although

$$F_{X_i}(t) = F_X(t) \qquad \text{for all } t, \quad i = 1, 2.$$

Thus, X_1 and X_2 would not be a random sample of a random variable. If, on the other hand, the sample is selected with replacement, then X_1 and X_2, as defined above, would be a random sample of the population random variable X.

In essentially all applications to be discussed in this book we shall assume that we are dealing with a random sample of a population random variable X. Our object, of course, is to use the observed sample values to make inferences about various aspects of the underlying population distribution. Thus the sampling is assumed to have been done in such a manner that Definition 2.2.1 is satisfied.

Example 2.2.3

Assume that telephone calls are placed to a large industrial switchboard, between 8 a.m. and 5 p.m. on Monday through Friday, like events in a Poisson process with parameter $\lambda = 20$ per hour, and that the number of calls are independent from day to day. Then, if X_i, $i = 1, 2, \ldots, 5$, are the number of incoming calls on Monday through Friday of a given week, respectively, $X_1, X_2, \ldots, X_5$ is a random sample of a Poisson random variable with parameter $9 \cdot 20 = 180$. The population sampled is conceptual and corresponds to the population of business days this firm will be in existence, at the current level of activity. Such questions as the number of incoming lines that should be adequate for the company to install can be investigated by making inferences about this conceptual population.

Example 2.2.4

Every item coming off a production line is either defective or not. Assume that the probability a given item is defective is p and that the occurrences of defectives are independent. If we then select a random sample of n items from the output of this production line and define

$$\begin{aligned} X_i &= 1 \qquad \text{if } i\text{th item is defective} \\ &= 0 \qquad \text{if not} \end{aligned}$$

$i = 1, 2, \ldots, n$, then $X_1, X_2, \ldots, X_n$ is a random sample of a Bernoulli random variable X with parameter p. The sample can be thought of as being taken from the population of the (conceptually) infinite number of items to be produced by this production line. The sample would be of use in making inferences about p, the probability of occurrence of a defective item.

Notice that if we have a random sample of a (scalar) random variable X, we can define the $n \times 1$ vector $\mathbf{X}$ which has components $X_1, X_2, \ldots, X_n$. Then the mean vector of $\mathbf{X}$ is $\boldsymbol{\mu}$, which has every component equal to μ_X, the mean of the population random variable, and the covariance matrix of $\mathbf{X}$ is $\sigma_X^2\mathbf{I}$, where σ_X^2 is the variance of the population random variable. Let us now discuss a second important definition.

DEFINITION 2.2.2. A *statistic* is any function of the elements of a random sample which does not depend on unknown parameters.

Statistics, then, are quantities whose observed values can be computed, once the sample has been taken. They form the bases of all statistical methods of making inferences about populations. Let us define some of the most commonly used statistics.

DEFINITION 2.2.3. Given $X_1, X_2, \ldots, X_n$ is a random sample of a random variable X

(i) The kth *sample moment* is

$$M_k = \frac{1}{n}\sum_{i=1}^{n} X_i^k, \qquad k = 1, 2, \ldots$$

(ii) The *sample mean*, $\bar{X}$, is the first sample moment:

$$\bar{X} = \frac{1}{n}\sum_{i=1}^{n} X_i.$$

(iii) The *sample variance* is

$$S^2 = \frac{1}{n-1}\sum_{i=1}^{n}(X_i - \bar{X})^2$$

and the *sample standard deviation* is

$$S = \sqrt{S^2}.$$

It is important to realize that, since statistics arc functions of random variables, they are themselvcs random variables. They will be denoted by capital letters and their observed values, after the sample is in hand, by lower case letters. The observed value of a statistic thus will vary from one sample to another; a major problem in mathematical statistics is to derive the distributions of useful statistics. We examine the distributions of certain statistics in the next section. It should also be observed that we can investigate vectors of statistics and may be interested in the joint distribution of several statistics, all defined as functions of the elements of the same sample.

EXERCISE 2.2

In each of the following $X_1, X_2, \ldots, X_n$ is a random sample of X.

1. Show that $E[M_k] = \eta_k$, the kth moment of X. Thus, the mean value of the distribution of M_k is η_k; this is interpreted to mean that over repeated samples the average value of M_k is η_k.

2. Show that $E[M_k^2] = (1/n)\eta_{2k} + [(n-1)/n]\eta_k^2$ and thus that the variance of M_k is

$$\frac{1}{n}[\eta_{2k} - \eta_k^2].$$

Note in particular, then, that $\bar{X}$ $(= M_1)$ has mean μ_X and variance σ_X^2/n. Thus the mean value of $\bar{X}$ is the same as that of X but its distribution is more concentrated about μ_X.

3. Show that $S^2 = [n/(n-1)][M_2 - \bar{X}^2]$ and thus $E[S^2] = \sigma_X^2$. It is because of this result that $n - 1$ is used as a divisor in S^2, rather than n.

4. As discussed above, let **X** be the $n \times 1$ vector with components $X_1, X_2, \ldots, X_n$. Show that

$$S^2 = \frac{1}{n-1}\mathbf{X}'\left(\mathbf{I} - \frac{1}{n}\mathbf{J}\right)\mathbf{X}, \qquad \bar{X} = \frac{1}{n}\mathbf{1}'\mathbf{X}$$

where $\mathbf{J} = \|1\|$, $\mathbf{1}$ is the $n \times 1$ vector with 1 in every position.

5. Show that $\mathbf{I} - (1/n)\mathbf{J}$ is an idempotent matrix. (**A** is idempotent if $\mathbf{A}^2 = \mathbf{A}$.) This type of matrix occurs very frequently in statistical applications.

6. A useful property of symmetric idempotent matrices is that every one of their characteristic roots is 0 or 1. Show this result to be true. (Hint: if **A** is idempotent and λ is a characteristic root of **A**, **x** the corresponding vector, then $\mathbf{A}^2\mathbf{x} = \mathbf{A}\mathbf{x} = \lambda\mathbf{x}$ and $\mathbf{x}'\mathbf{A}'\mathbf{A}\mathbf{x} = \mathbf{x}'\mathbf{A}^2\mathbf{x} = \mathbf{x}'\mathbf{A}\mathbf{x}$.)

7. Since the trace of a symmetric matrix **A**, tr(**A**), gives the sum of its characteristic roots, show that if **A** is symmetric, idempotent, then

$$\mathrm{tr}(\mathbf{A}) = r(\mathbf{A}) \qquad \text{(rank of } \mathbf{A}).$$

8. If **A** is symmetric, idempotent, show that

$$r(\mathbf{A}) = \sum_i \sum_j a_{ij}^2.$$

9. Express the moment generating function for $\bar{X} = (1/n)\sum_{i=1}^n X_i$ in terms of $M_X(t)$.

10. If X is normal, μ_X, σ_X, what is the distribution for $\bar{X}$?

11. If X is Poisson with parameter λ, what is the probability function for $\bar{X}$?

12. If X is exponential with parameter λ, what is the density function for $\bar{X}$?

13. If X is a gamma random variable with parameters r and λ, what is the density function for $\bar{X}$?

14. If X is binomial with parameters m and p, what is the probability function for $\bar{X}$?

15. Assume that $X \sim N(0, \sigma)$, $n = 4$. What is the probability that at least 3 of the sample values exceed 0? that the largest exceeds 0?

16. Let $X_1, X_2, \ldots, X_n$ be a random sample of a normal random variable with mean μ and variance σ^2; if $\mathbf{X}' = (X_1, X_2, \ldots, X_n)$, what is the distribution for **X**?

17. Assume **X** is multivariate normal with mean $\boldsymbol{\mu}$ and covariance matrix $\boldsymbol{\Sigma}$; $\mathbf{X}_i$, $i = 1, 2, \ldots, n$ is a random sample of **X**. Define

$$\bar{\mathbf{X}} = \frac{1}{n}\sum_{i=1}^n \mathbf{X}_i;$$

What is the distribution for $\bar{\mathbf{X}}$?

2.3 χ^2, t, and F Random Variables

In this section we shall derive the distributions of several random variables which frequently occur in problems where statistical methods are used. The first of these that we shall study is called the χ^2 random variable. It occurs quite naturally when sampling from a normal population. Assume then that X is a normal random variable with mean 0 and variance σ^2 and define $Y = X^2/\sigma^2$ (note that X/σ is $N(0, 1)$). Then the range of Y is clearly $R_Y = \{y: \ y > 0\}$ and

$$f_Y(y) = \frac{\sigma}{2\sqrt{y}}\,[f_X(\sigma\sqrt{y}) + f_X(-\sigma\sqrt{y})]$$

$$= \frac{1}{\sqrt{2\pi}}\frac{1}{\sqrt{y}}\,e^{-y/2} = \frac{1}{\Gamma(1/2)2^{1/2}}\,y^{-1/2}e^{-y/2}$$

(see Exercise 1.3.18). Notice, then, from Section 1.3, that Y is a gamma random variable with parameters $r = 1/2$, $\lambda = 1/2$. It is also called a (central) χ^2 *random variable* with one degree of freedom (because it is the square of a single standard normal random variable). Thus, since it is a special gamma random variable, its moment generating function is

$$m_Y(t) = \frac{1}{[1 - t/(1/2)]^{1/2}} = \frac{1}{(1 - 2t)^{1/2}}$$

Then, suppose $X_1, X_2, \ldots, X_n$ is a random sample of a normal random variable with mean 0 and variance σ^2 and define

$$Y_i = \frac{X_i^2}{\sigma^2}, \qquad i = 1, 2, \ldots, n$$

$$W = \sum_{i=1}^{n} Y_i.$$

Then $Y_1, Y_2, \ldots, Y_n$ are independent, each χ^2 random variables with one degree of freedom and the moment generating function for W is

$$m_W(t) = \prod_{i=1}^{n} m_{Y_i}(t)$$

$$= \prod_{i=1}^{n}\frac{1}{(1 - 2t)^{1/2}} = \frac{1}{(1 - 2t)^{n/2}},$$

so W is a gamma random variable with parameters $r = n/2$, $\lambda = 1/2$, and its density is

$$f_W(w) = \frac{1}{\Gamma\left(\frac{n}{2}\right)}\frac{1}{2^{n/2}}\,w^{(n/2)-1}e^{-w/2};$$

W is called the (central) χ^2 random variable with n degrees of freedom (since it is the sum of squares of n independent standard normal random variables). We shall denote a generic random variable with this distribution by χ^2_n. Note that $\mu_{\chi_n^2} = n$, $\sigma^2_{\chi_n^2} = 2n$, and the maximum of the density function occurs at $w = n - 2$, for $n \geq 2$ (this is called the *modal value* for χ^2_n).

Example 2.3.1

Assume that a single round is fired from a rifle at a bullseye target and that the coordinates of the point at which it hits the target are (X_1, X_2) (the origin is at the center of the target) where X_1 and X_2 are independent standard normal random variables. Then the radial distance between the target center and the impact point is $\sqrt{X_1^2 + X_2^2} = \sqrt{W}$, where W is a χ^2 random variable with 2 degrees of freedom (see Figure 2.3.1). If the radius of the inner ring is 1/8 unit, then the probability that the shot will land somewhere

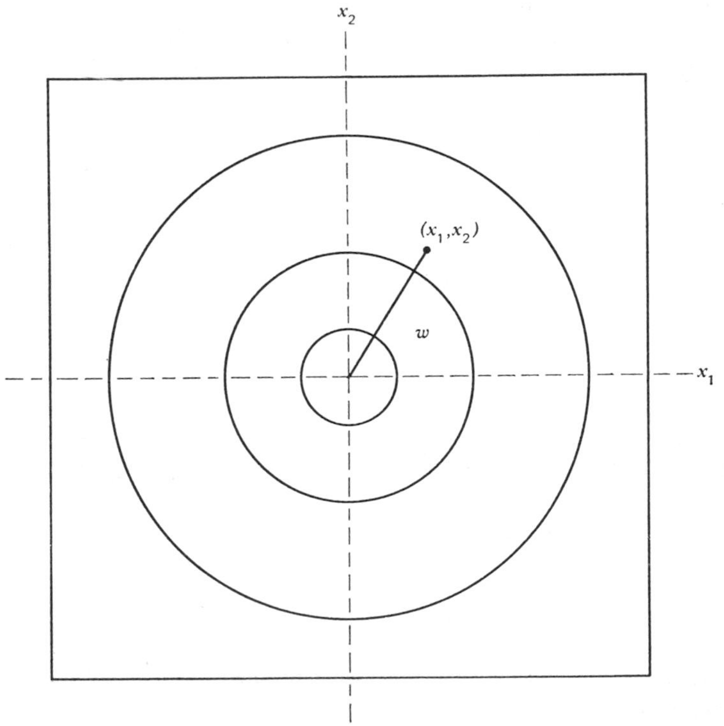

FIGURE 2.3.1. Target and coordinate system.

inside that ring is

$$P\left(\sqrt{W} < \frac{1}{8}\right) = P\left(W < \frac{1}{64}\right) = \int_0^{1/64} \frac{1}{2} e^{-y/2}\, dy = 1 - e^{-1/128} = .0078.$$

Table 2 presents a table of the χ_n^2 distribution function for the degrees of freedom from 1 to 30. We shall use $\chi_n^2(\alpha)$ to denote the percentiles of the central χ_n^2 distribution with n degrees of freedom. Thus, $P(\chi_n^2 \leq \chi_n^2(\alpha)) = \alpha$. For degrees of freedom $n > 30$, the χ_n^2 distribution function is well approximated by assuming that $\sqrt{2W} - \sqrt{2n - 1}$ is a standard normal random variable, where W is χ_n^2; thus for $n > 30$,

$$\begin{aligned}
\alpha &= P(\chi_n^2 \leq \chi_n^2(\alpha)) \\
&= P(\sqrt{2W} - \sqrt{2n - 1} \leq \sqrt{2\chi_n^2(\alpha)} - \sqrt{2n - 1}) \\
&\doteq P(Z \leq z(\alpha))
\end{aligned}$$

where $z(\alpha)$ is a percentile of the standard normal distribution. Note then, for $n > 30$,

$$\chi_n^2(\alpha) \doteq \frac{1}{2}(z(\alpha) + \sqrt{2n - 1})^2.$$

By the central limit theorem we would certainly expect W to be approximately normally distributed, since it is the sum of independent, identically distributed random variables. The above approximation, derived by Fisher, is suggested because the distribution of $\sqrt{2W} - \sqrt{2n - 1}$ is better approximated by the standard normal, for any given n, than is the distribution of $(1/\sqrt{2n})(W - n)$.

A related distribution, which also proves useful in problems of inference, is that of the *noncentral* χ^2. The noncentral χ^2 random variable with 1 degree of freedom, which we shall denote $\chi_1'^2$ is the square of a normal random variable whose variance is 1, as before, but whose mean is $\mu \neq 0$. Accordingly, assume X is $N(\mu, \sigma)$ and let $V = (X^2/\sigma^2)$; then, as before

$$\begin{aligned}
f_V(v) &= \frac{\sigma}{2\sqrt{v}}[f_X(\sigma\sqrt{v}) + f_X(-\sigma\sqrt{v})] \\
&= \frac{1}{2\sqrt{v}\sqrt{2\pi}}[e^{-\frac{1}{2}(\sqrt{v}-\mu/\sigma)^2} + e^{-\frac{1}{2}(\sqrt{v}+\mu/\sigma))^2}] \\
&= \frac{e^{-(\frac{1}{2})(v+\mu^2/\sigma^2)}}{2\sqrt{v}\sqrt{2\pi}}\left[e^{\frac{\mu}{\sigma}\sqrt{v}} + e^{-\frac{\mu}{\sigma}\sqrt{v}}\right].
\end{aligned}$$

Note, then, if we expand each of the two exponentials in a Taylor series, and add them together, the odd powers cancel each other and we have

$$f_V(v) = \frac{e^{-\frac{1}{2}(v+\mu^2/\sigma^2)}}{\sqrt{v}\sqrt{2\pi}} \sum_{j=0}^{\infty} \frac{(\mu^2 v)^j}{(2j)!} \left(\frac{1}{\sigma^2}\right)^j.$$

The moment generating function for V (see problem 4 below) is

$$M_V(t) = \frac{e^{t\mu^2/[(1-2t)\sigma^2]}}{(1-2t)^{\frac{1}{2}}}.$$

Notice that the density of V involves (μ^2/σ^2). (μ^2/σ^2) is called the *noncentrality parameter* of V.

If $X_1, X_2, \ldots, X_n$ is a random sample of a normal random variable X with mean μ and variance σ^2, let

$$U = \sum_{i=1}^{n} \frac{X_i^2}{\sigma^2} = \sum_{i=1}^{n} V_i.$$

Then

$$M_U(t) = \prod_{i=1}^{n} M_{V_i}(t) = \frac{e^{nt\mu^2/[(1-2t)\sigma^2]}}{(1-2t)^{n/2}}$$

and U is the *noncentral* χ^2 random variable with n degrees of freedom and noncentrality parameter $n\mu^2/\sigma^2$. It is easily verified that $E[U] = n + n\mu^2/\sigma^2$, $\sigma_U^2 = 2n + 4n\mu^2/\sigma^2$. Note that, if $\mu = 0$, the noncentral χ^2 random variable reduces to a central χ^2, as of course it should. The noncentral χ^2 distribution will occur when we examine the power functions for certain tests of hypotheses in Chapter 5.

Let us now examine two more basic distributions, which occur as functions of samples from normal populations and which, like the χ^2, will prove very useful in our studies of certain problems of statistical inference. The first of these is the *t distribution;* we shall use a capital T to denote this random variable but, because of common usage the distribution will be referred to as that of a t random variable. Assume that Z is a standard normal random variable, W is an independent χ^2 random variable with n degrees of freedom and define

$$T = \frac{Z}{\sqrt{W/n}};$$

this random variable has the *t distribution* with n degrees of freedom. Its density function can be derived by transforming the joint density of W and Z as follows: Since W and Z are independent,

$$f_{W,Z}(w, z) = f_W(w) f_Z(z) = \frac{1}{\Gamma(n/2)} \frac{1}{2^{(n/2)}} w^{(n/2)-1} e^{-w/2} \cdot \frac{1}{\sqrt{2\pi}} e^{-z^2/2}.$$

Now, let $t = z/\sqrt{(w/n)} = \sqrt{n}(z/\sqrt{w})$, $u = w$; then the reciprocal of the jacobian of the transformation is

$$\frac{1}{J} = \begin{vmatrix} \dfrac{\partial t}{\partial z} & \dfrac{\partial t}{\partial w} \\ \dfrac{\partial u}{\partial z} & \dfrac{\partial u}{\partial w} \end{vmatrix} = \begin{vmatrix} \sqrt{\dfrac{n}{w}} & -\dfrac{1}{2}\sqrt{n}\,\dfrac{z}{w\sqrt{w}} \\ 0 & 1 \end{vmatrix} = \sqrt{\frac{n}{w}}$$

and thus $J = \sqrt{(w/n)} = \sqrt{(u/n)}$. It is clear that the ranges of the new variables are $-\infty < t < \infty$, $u > 0$ and the joint density of T and U is

$$\begin{aligned} f_{T,U}(t, u) &= f_{W,Z}(u, t\sqrt{u/n})\,|J| \\ &= \frac{1}{\sqrt{2\pi}\,\Gamma\left(\frac{n}{2}\right)2^{n/2}} u^{(n/2)-1}e^{-u/2}e^{-(t^2u)/(2n)}\sqrt{u/n} \\ &= \frac{1}{\sqrt{n}\sqrt{2\pi}\,\Gamma\left(\frac{n}{2}\right)2^{n/2}} u^{(n-1)/2}e^{-(u/2)[1+(t^2/n)]}. \end{aligned}$$

The marginal density function for T then is

$$f_T(t) = \int_0^\infty \frac{1}{\sqrt{n}\sqrt{2\pi}\,\Gamma\left(\frac{n}{2}\right)2^{n/2}} u^{(n-1)/2}e^{-(u/2)[1+(t^2/n)]}\,du.$$

In this integral let $y = (u/2)[1 + (t^2/n)]$, $dy = (du/2)[1 + (t^2/n)]$ and we have

$$\begin{aligned} f_T(t) &= \int_0^\infty \frac{1}{\sqrt{n}\sqrt{2\pi}\,\Gamma\left(\frac{n}{2}\right)2^{n/2}}\left[\frac{2y}{1 + (t^2/n)}\right]^{(n-1)/2} e^{-y}\,\frac{2\,dy}{1 + (t^2/n)} \\ &= \frac{2^{(n+1)/2}}{\sqrt{n}\sqrt{2\pi}\,\Gamma(n/2)2^{n/2}}\,\frac{1}{[1 + (t^2/n)]^{(n+1)/2}}\int_0^\infty y^{(n-1)/2}e^{-y}\,dy \\ &= \frac{\Gamma[(n+1)/2]}{\sqrt{\pi n}\,\Gamma(n/2)}\,\frac{1}{(1 + (t^2/n))^{(n+1)/2}}. \end{aligned}$$

This is the density of the t distribution with n degrees of freedom. Notice that the only parameter is the degrees of freedom, n, from the denominator χ^2 random variable. Table 3 presents selected percentiles of the t distribution for degrees of freedom up to 30. As with the χ^2 random variable, we shall let $t_n(\alpha)$ be defined by

$$P(T \leq t_n(\alpha)) = \alpha.$$

For $n > 30$, the distribution function for T is fairly well approximated by the standard normal distribution function and thus $t_n(\alpha) = z(\alpha)$ for $n > 30$. Notice that the t density function is symmetric about the line $t = 0$ for all n and thus the modal t value is 0.

As with the χ_n^2 random variable, we shall find use in Chapter 5 for a noncentral form for the t distribution. This random variable occurs as follows: Assume that X is $N(\mu, 1)$, $\mu \neq 0$, and V is χ_n^2, X and V are independent. Define

$$T' = \frac{X}{\sqrt{V/n}} = \sqrt{n}\,\frac{X}{\sqrt{V}};$$

then T' has the noncentral t distribution with n degrees of freedom and noncentrality parameter $(\mu\sqrt{n})/\sigma = \eta$ (denoted as the $t'_{n,\eta}$ distribution). Note then that

$$\frac{X - \mu}{\sqrt{V/n}} = T' - \frac{\mu}{\sqrt{V/n}} = T$$

and thus $T' = T + (\mu\sqrt{n})/\sqrt{V}$. The distribution function for T' has been tabled by Lieberman and Resnikoff for various values of the parameters.

Another important type of random variable which occurs frequently in problems of statistical inference is the *F random variable*. It is defined as follows: Assume that U is a (central) χ^2 with g degrees of freedom, V is a (central) χ^2 with h degrees of freedom and U and V are independent. Then

$$F = \frac{U}{g}\bigg/\frac{V}{h} = \frac{Uh}{Vg}$$

has the *F distribution* with g and h degrees of freedom. Let us now derive its density function. We have

$$f_{U,V}(u, v) = f_U(u)f_V(v) = \frac{1}{\Gamma(g/2)2^{g/2}}\,u^{(g/2)-1}e^{-(u/2)}\,\frac{1}{\Gamma(h/2)2^{h/2}}\,v^{(h/2)-1}e^{-v/2};$$

make the change of variable $f = uh/vg$, $w = v$ and the reciprocal of the jacobian is

$$\frac{1}{J} = \begin{vmatrix} \dfrac{\partial f}{\partial u} & \dfrac{\partial f}{\partial v} \\[2ex] \dfrac{\partial w}{\partial u} & \dfrac{\partial w}{\partial v} \end{vmatrix} = \begin{vmatrix} \dfrac{h}{vg} & -\dfrac{uh}{v^2 g} \\[2ex] 0 & 1 \end{vmatrix} = \frac{h}{vg} = \frac{h}{wg}.$$

The ranges of f and w are $f > 0$, $w > 0$. Then the joint density of F and W is

$$f_{F,W}(f, w) = f_{U,V}\left(\frac{fwg}{h}, w\right) |J|$$

$$= \frac{1}{\Gamma(g/2)\Gamma(h/2)2^{(g+h)/2}}\left(\frac{fwg}{h}\right)^{(g/2)-1} w^{(h/2)-1}e^{-w/2[1+(fg/h)]}\frac{gw}{h},$$

and the marginal density for F is

$$f_F(f) = \frac{(g/h)^{g/2}}{\Gamma(g/2)\Gamma(h/2)2^{(g+h)/2}} f^{(g/2)-1}\int_0^\infty w^{(g+h)/2-1}e^{-(w/2)[1+(fg/h)]}dw;$$

letting $x = (w/2)[1 + (fg/h)]$, $dx = (dw/2)[1 + (fg/h)]$, we have

$$f_F(f) = \frac{(g/h)^{g/2}}{\Gamma(g/2)\Gamma(h/2)2^{(g+h)/2}} f^{(g/2)-1}[1 + (fg/h)]^{-(g+h)/2}2^{(g+h)/2}$$

$$\times \int_0^\infty x^{[(g+h)/2]-1}e^{-x}dx$$

$$= \frac{\Gamma[(g+h)/2]}{\Gamma(g/2)\Gamma(h/2)}\frac{(g/h)^{g/2}f^{(g/2)-1}}{[1 + (fg/h)]^{(g+h)/2}}.$$

Table 4 in the Appendix gives some values from the distribution function of an F random variable with various values for the degrees of freedom g and h. (We denote a generic random variable with this distribution by $F_{g,h}$.) We shall define $F_{g,h}(\alpha)$ to be the value of $F_{g,h}$ such that

$$P(F_{g,h} \le F_{g,h}(\alpha)) = \alpha.$$

Notice that an $F_{g,h}$ random variable is the ratio of 2 independent χ^2 random variables, each divided by their degrees of freedom; thus $1/(F_{g,h})$ is also such a ratio and again has an $F_{h,g}$ distribution with the denominator and numerator degrees of freedom interchanged; thus

$$\alpha = P(F_{g,h} \le F_{g,h}(\alpha)) = P\left(\frac{1}{F_{g,h}} \ge \frac{1}{F_{g,h}(\alpha)}\right)$$

$$= 1 - P\left(\frac{1}{F_{g,h}} \le \frac{1}{F_{g,h}(\alpha)}\right) = 1 - P\left(F_{h,g} \le \frac{1}{F_{g,h}(\alpha)}\right)$$

and therefore

$$F_{h,g}(1 - \alpha) = F_{g,h}^{-1}(\alpha).$$

An F random variable is a transform of a beta random variable, as can be verified by letting $x = (fg/h)/[1 + (fg/h)]$ in the F density function; X is then a beta random variable with parameters $\alpha = g/2$, $\beta = h/2$ (see problem 5 below).

It will be recalled that a t random variable with f degrees of freedom is the ratio of a standard normal random variable over the square root of an independent χ_f^2 random variable over its degrees of freedom; also, that the square of a standard normal random variable is a χ_1^2 random variable. Thus, t^2 is the ratio of 2 independent χ^2 random variables, each divided by its degrees of freedom and therefore is an F random variable with 1 and f degrees of freedom. Thus, as can be verified from Tables 3 and 4,

$$t_f^2(\alpha) = F_{1,f}(2\alpha - 1),$$

$$t_f\left(\frac{\alpha + 1}{2}\right) = \sqrt{F_{1,f}(\alpha)}.$$

Let us now turn our attention to a very important result regarding samples selected from a normal population. The sample mean, $\bar{X}$, and the sample variance, S^2, were introduced in the last section. If we have a random sample from a normal population, then these two statistics are independently distributed, even though they are both functions of the same sample values. This result is contained in the following theorem.

THEOREM 2.3.1. $X_1, X_2, \ldots, X_n$ is a random sample of a random variable X which is $N(\mu, \sigma)$. Then

$$\bar{X} = \frac{1}{n}\sum_{i=1}^{n} X_i \quad \text{is} \quad N\left(\mu, \frac{\sigma}{\sqrt{n}}\right),$$

$$\frac{(n-1)S^2}{\sigma^2} = \frac{\sum (X_i - \bar{X})^2}{\sigma^2} \quad \text{is} \quad \chi_{n-1}^2$$

and $\bar{X}$ and S^2 are independent random variables.

Proof: Let $\mathbf{X}$ be the $n \times 1$ vector having the sample random variables as its components. Then $\mathbf{X}$ is multivariate normal with $\boldsymbol{\mu} = \mu\mathbf{1}$, $\boldsymbol{\Sigma} = \sigma^2\mathbf{I}$ (see Exercise 2.2.16). Let $\mathbf{C}$ be an orthogonal matrix with first row $(1/\sqrt{n})\mathbf{1}' = \mathbf{c}_1$ and transform to $\mathbf{Y} = \mathbf{CX}$. Then $\mathbf{Y}$ is multivariate normal with mean $\mathbf{C}\boldsymbol{\mu}$ and covariance matrix $\sigma^2\mathbf{I}$. Note that

$$Y_1 = \mathbf{c}_1\mathbf{X} = \frac{1}{\sqrt{n}}\mathbf{1}'\mathbf{X} = \sqrt{n}\,\bar{X}$$

and

$$E[Y_1] = \mathbf{c}_1\boldsymbol{\mu} = \left(\frac{1}{\sqrt{n}}\mathbf{1}'\right)(\mu\mathbf{1}) = \sqrt{n}\,\mu;$$

$$E[Y_j] = \mathbf{c}_j\boldsymbol{\mu} = \mathbf{c}_j\mu\mathbf{1} = \mu\sqrt{n}\,\mathbf{c}_j\mathbf{c}_1' = 0, \qquad j = 2, \ldots, n,$$

so $Y_2, Y_3, \ldots, Y_n$ are independent, each with mean 0, variance σ^2, and independent of Y_1. Furthermore, since $\mathbf{C}$ is orthogonal,

$$\sum_{i=1}^{n} X_i^2 = \mathbf{X'X} = \mathbf{Y'CC'Y} = \sum_{i=1}^{n} Y_i^2$$

and $(n-1)S^2 = \sum_{i=1}^n X_i^2 - n\bar{X}^2 = \sum_{i=1}^n Y_i^2 - Y_1^2 = \sum_{i=2}^n Y_i^2$. Thus $\bar{X}$ is a function only of Y_1, S^2 is a function only of $Y_2, Y_3, \ldots, Y_n$ and, since the Y_i's are independent, $\bar{X}$ and S^2 are independent. Also, $\bar{X} = (1/\sqrt{n})Y_1$ is $N(\mu, \sigma/\sqrt{n})$, and $[(n-1)S^2]/\sigma^2 = \sum_{i=2}^n (Y_i^2/\sigma^2)$ is a χ^2_{n-1} random variable.

Another important result is contained in the following theorem.

THEOREM 2.3.2. If $X_1, X_2, \ldots, X_n$ is a random sample of a random variable X which is $N(\mu, \sigma)$ then $(\bar{X}-\mu)/(S/\sqrt{n})$ has the t_f distribution with $f = n-1$ degrees of freedom.

Proof: $(\bar{X}-\mu)/(\sigma/\sqrt{n})$ is $N(0, 1)$ and $[(n-1)S^2]/\sigma^2$ is χ^2_{n-1}, and $\bar{X}$ and S^2 are independent, from Theorem 2.3.1. Thus

$$\frac{\bar{X}-\mu}{\sigma/\sqrt{n}} \Bigg/ \sqrt{\frac{(n-1)S^2}{\sigma^2} \Big/ (n-1)} = \frac{\bar{X}-\mu}{S/\sqrt{n}}$$

is the ratio of a standard normal random variable divided by the square root of an independent χ^2_{n-1} random variable over its degrees of freedom. As we saw above, the ratio then has the t distribution with $n-1$ degrees of freedom.

The remarkable thing about this result is that the standard deviation σ cancels in forming the ratio and the t distribution has only the single known parameter, $n-1$. As we shall see, this fact leads to some very important results in statistical inference.

Example 2.3.2

The above result can be used to compute exact probabilities about the distance between $\bar{X}$ and μ, when sampling from a normal population (some other, more important implications will be mentioned in Chapters 4 and 5). From Appendix Table 3 of the t distribution, we see that for 9 degrees of freedom $t_9(.95) = 1.833$. Thus, if we have a random sample of 10 observations from a normal population,

$$P\left(\frac{\bar{X}-\mu}{S/\sqrt{10}} \le 1.833\right) = .95$$

and

$$P\left(\frac{|\bar{X}-\mu|}{S/\sqrt{10}} \le 1.833\right) = .9$$

because of the symmetry of the t distribution. Note then that the probability is .9 that $|\bar{X} - \mu|$ is no larger than $(S/\sqrt{10})1.833 = .580S$. Thus in 90% of the samples of size 10 selected from a normal population, $\bar{X}$ differs from μ by no more than $.580S$, where S is the sample standard deviation from the same sample. Given the results of the sampling, $.580S$ can be readily computed.

EXERCISE 2.3

1. Assume that the thickness of the bark on 20-year-old cork trees, at a height of 4 feet, is normally distributed with $\mu = 3.5$ inches and $\sigma = .75$ inch. What is the probability that the sample average, $\bar{X}$, of the thicknesses measured on 9 trees will exceed 3.8 inches? that it will be between 3.4 and 3.6 inches?

2. For the situation described in question 1, what is the probability that the sample variance exceeds 1 square inch?

3. Sketch the t density function for $f = 4$, 6, and 9 degrees of freedom.

4. Verify that the moment generating function for a χ'^2_1 random variable is

$$M_{\chi'^2_1}(t) = \frac{e^{t\mu^2/(1-2t)\sigma^2}}{(1-2t)^{1/2}}.$$

5. Show that $V = (g/h)F_{g,h}/(1 + (g/h)F_{g,h})$ is a beta random variable, where $F_{g,h}$ has the F distribution with g and h degrees of freedom.

6. Assume that a dart is thrown at a target and that the coordinates of the place it impacts are (X_1, X_2), where the origin of the coordinate system is at the center of the target. Let $\mathbf{X}' = (X_1, X_2)$ and assume $\mathbf{X}$ is bivariate normal, $\boldsymbol{\mu}$, $\boldsymbol{\Sigma} = \mathbf{I}$. What is the distribution of $R^2 = \mathbf{X}'\mathbf{X}$, the square of the radial distance between $\mathbf{0}$ and $\mathbf{X}$?

7. For problem 6, find the density of R, the radial distance itself.

8. Show that the χ^2 distribution reproduces itself; i.e., if X_i is $\chi^2_{f_i}$, $i = 1, 2, \ldots, n$ and the X_i's are independent, show that

$$Y = \sum_{i=1}^{n} X_i$$

is χ^2_f with degrees of freedom $f = \sum_{i=1}^{n} f_i$.

9. Show that the noncentral χ^2 distribution reproduces itself.

10. If $X_1, X_2, \ldots, X_n$ is a sample of a normal random variable with mean μ, variance σ^2, show that

$$E[S] \neq \sigma.$$

11. Assume that the lifetime of a certain type of electron tube, in a given usage, is an exponential random variable with parameter λ; thus

$$f_X(x) = \lambda e^{-\lambda x}.$$

If $X_1, X_2, \ldots, X_n$ is a random sample of n such lifetimes, what is the distribution of $2\lambda n\bar{X}$? If $n = 5$, $\lambda = 1$, what is the probability that $\bar{X} < .9$?

12. Compute the mean and variance of an $F_{g,h}$ random variable.

13. Given a random sample of size n from a normal population, what is the variance of S^2? of S?

14. Formally derive the density of

$$Y = T_f^2$$

where T_f has the t distribution with f degrees of freedom, from the density of T_f.

15. Assume that $\mathbf{X}$ is $p \times 1$, multivariate normal with mean vector $\mathbf{0}$ and covariance matrix $\mathbf{\Sigma}$. Show that $\mathbf{X}'\mathbf{\Sigma}^{-1}\mathbf{X}$ is a χ^2 random variable with p degrees of freedom. (Hint: In Section 1.4 we saw that there exists a non-singular matrix $\mathbf{C}$ such that $\mathbf{C\Sigma C}' = \mathbf{I}$; use this matrix to transform to $\mathbf{Y} = \mathbf{CX}$.)

16. Assume that $\mathbf{X}$ is p-variate normal with mean $\boldsymbol{\mu}$ and covariance $\mathbf{\Sigma}$ and show that

$$W = (\mathbf{X} - \boldsymbol{\mu})'\mathbf{\Sigma}^{-1}(\mathbf{X} - \boldsymbol{\mu})$$

is a χ_p^2 random variable.

17. Assume $\mathbf{X}$ is as defined in problem 16 and let $Y = \mathbf{1}'\mathbf{X}$, where $\mathbf{1}$ is the $p \times 1$ vector with 1 in every position. Show that

$$\frac{\mathbf{1}'(\mathbf{X} - \boldsymbol{\mu})(\mathbf{X} - \boldsymbol{\mu})'\mathbf{1}}{\mathbf{1}'\mathbf{\Sigma}\mathbf{1}}$$

is $\chi^2(1)$, by examining the distribution of Y.

CHAPTER 3

Estimation

3.1 Introduction

We shall in this chapter begin our study of statistical inference, trying to derive facts concerning a population by examining only a random sample from the population. This chapter is concerned with the problem of *point estimation*; it is assumed that the form of the population distribution is known (normal, binomial, Poisson, etc.) but that one or more parameters of the population distribution are unknown. The problem then is, given a random sample of observed values selected from the population, what function(s) of these numbers might be expected to provide a good numerical guess (point estimate) of the value(s) of the unknown parameter(s)? Indeed, what is a good criterion for judging point estimates and for choosing between different possible alternative methods of computing point estimates?

For example, on a given election day a certain (unknown) number of people will actually vote, and, among these, an (unknown) proportion p will vote in a given way. Let p^* be the proportion of eligible voters that say they will vote in the specified way one week prior to the election. If it is assumed that $p^* = p$, essentially, then a sample of eligible voters might be selected one week prior to the election and each asked which way he will vote. The responses received should certainly contain information about p^* (and therefore p). Exactly what should be done with the observed responses to best estimate p^*, in some sense?

As a second example, well accepted physical "laws" dictate that the number of particles emitted from a radioactive pile should behave like events in a Poisson process. However, they do not uniquely specify the value of the rate parameter λ, for a given set of environmental conditions. Thus if we were to observe X, the number of particles emitted during a fixed period of time of length t, it is assumed that X is Poisson with parameter λt, and given a random sample of values of X, it then would be of interest to use the observed values, in some way, to estimate the parameter λ.

Indeed, it might prove of great interest to estimate the manner in which λ changes as the environmental conditions of the pile are changed. Many other examples could be given.

We shall rigorously adhere to the notation of using capital letters for random variables and lower case letters for their observed values. The *estimator* for an unknown parameter is a function of $X_1, X_2, \ldots, X_n$, the elements of the random sample, and as such is itself a random variable; thus estimators will be denoted by capital letters, frequently the capital letter of the parameter being estimated. The observed value of the estimator, computed from a given set of sample values, is called the *estimate* of the parameter and will be denoted by a lower case letter. An estimate is a number, certainly not a random variable. Neither of these quantities should be confused with the value of the parameter itself. Thus, if we are interested in estimating the parameter λ, we might let λ represent the *parameter*, $\hat{\Lambda}$ (capital lamda hat) represent the *estimator*, and $\hat{\lambda}$ represent its observed value for our given sample, the *estimate*. It would seem ideal, of course, if we could in some manner make $|\hat{\lambda} - \lambda|$ very small, or even 0, for our sample; because of the inherent randomness, however, this ideal goal cannot be achieved with certainty, in general.

3.2 Method of Moments and Maximum Likelihood

The oldest general criterion for generating estimates of unknown parameters from a sample, proposed by K. Pearson about 1891, is the *method of moments*. It is generally applicable and gives a fairly simple method for determining the estimates in most cases; however, as we shall see, it does not in general necessarily give rise to estimators which are as good, in certain senses, as those generated by the *method of maximum likelihood*, first proposed by R. A. Fisher about 1912. Thus, in cases in which the two methods give rise to different estimates, generally the maximum likelihood estimate is to be preferred.

The *method of moments* estimates, given a sample, are determined as follows: Assume we are given a population and that the distribution of the population random variable X is dependent upon the values of k unknown parameters $\theta_1, \theta_2, \ldots, \theta_k$. Generally, the first k population moments will depend on these parameters. (If not the first k, then some set of k surely will); i.e.,

$$E[X^i] = g_i(\theta_1, \theta_2, \ldots, \theta_k), \qquad i = 1, 2, \ldots, k.$$

The first k (observed) sample moments, $m_1, m_2, \ldots, m_k$ are equated to these population moments, generating k equations in k unknowns: $m_i = g_i(\theta_1, \theta_2, \ldots, \theta_k)$, $i = 1, 2, \ldots, k$. The method of moments estimates are the solutions, $\tilde{\theta}_i = h_i(m_1, m_2, \ldots, m_k)$, of these equations. The method of moments *estimators* are the same functions of the sample moments, M_1, $M_2, \ldots, M_k$; i.e., $\tilde{\Theta}_i = h_i(M_1, M_2, \ldots, M_k)$. These quantities are random variables whose behavior can be examined from one sample to another.

Example 3.2.1

Suppose $X_1, X_2, \ldots, X_n$ is a random sample from a normal population with mean μ and variance σ^2. Then the first two population moments are

$$\begin{aligned} E[X] &= \mu &&= g_1(\mu, \sigma^2) \\ E[X^2] &= \mu^2 + \sigma^2 &&= g_2(\mu, \sigma^2). \end{aligned}$$

Thus, the equations to be solved to determine the method of moments estimates are

$$\mu = m_1 = \bar{x}$$

$$\mu^2 + \sigma^2 = m_2 = \frac{1}{n}\sum_{i=1}^{n} x_i^2.$$

The solutions are easily found to be $\tilde{\mu} = m_1 = \bar{x} = h_1(m_1, m_2)$

$$\tilde{\sigma}^2 = m_2 - m_1^2 = \frac{1}{n}\sum_{i=1}^{n} x_i^2 - \bar{x}^2 = \frac{1}{n}\sum_{i=1}^{n} (x_i - \bar{x})^2 = h_2(m_1, m_2).$$

The corresponding estimators for μ and σ^2 then are $M_1 = \bar{X}$ and $M_2 - M_1^2 = [(n-1)/n]S^2$, respectively. Notice that if we had used μ and σ as the parameters, rather than μ and σ^2, then the estimators would be $\bar{X}$ and $S\sqrt{(n-1)/n}$, respectively.

Example 3.2.2

Assume $X_1, X_2, \ldots, X_n$ is a random sample of a uniform random variable on the interval (a, b). Then the first two population moments are

$$E[X] = \frac{b + a}{2} = g_1(a, b)$$

$$E[X^2] = \frac{b^2 + ab + a^2}{3} = g_2(a, b)$$

and the equations to be solved to determine the estimates of a and b are

$$m_1 = \frac{b + a}{2}$$

$$m_2 = \frac{b^2 + ab + a^2}{3};$$

the solutions are readily found to be

$$\tilde{b} = m_1 + \sqrt{3(m_2 - m_1^2)} = \bar{x} + s\sqrt{\frac{3(n-1)}{n}}$$

$$\tilde{a} = m_1 - \sqrt{3(m_2 - m_1^2)} = \bar{x} - s\sqrt{\frac{3(n-1)}{n}}$$

where s^2 is the observed value of the sample variance. The estimators then are $\tilde{B} = \bar{X} + S\sqrt{[3(n-1)/n]}$, $\tilde{A} = \bar{X} - S\sqrt{[3(n-1)/n]}$. An interesting fact about these estimators can be observed by looking at a particular set of sample values. Suppose $n = 9$ and we observe .5, .6, .1, 1.3, .9, 1.6, .7, .9, 1.0 as the sample values. Then, since we are assuming each of these is an observed value of a uniform (a, b) random variable, it is plain that $a \leq .1$ and $b \geq 1.6$, since these values in particular must lie in (a, b). It is easily found that, for this sample, $\bar{x} = .84$, $m_2 = .89$, $\sqrt{3(m_2 - \bar{x}^2)} = .744$ and thus $\tilde{a} = .096$, $\tilde{b} = 1.584$; thus the estimated $(\tilde{a}, \tilde{b})$ interval does not include all the observed sample values. Thus we can see that an intuitively appealing procedure such as the method of moments can give rise to sample estimates which are obviously defective.

The *method of maximum likelihood* is based upon a different rationale than is the method of moments. Rather than equating sample moments to population moments, it sets the estimate equal to the value of the unknown parameter which maximizes the probability of the observed sample results. This, in a sense, gives that estimate which is the "most likely" value for the unknown parameter, based upon the observed sample. To see the details of how such estimators are defined we must first define the likelihood function for the sample.

DEFINITION 3.2.1. Assume $X_1, X_2, \ldots, X_n$ is a random sample from a population whose distribution depends on the values of k unknown parameters, $\theta_1, \theta_2, \ldots, \theta_k$. The *likelihood function* of the sample is defined to be

$$L(\theta_1, \theta_2, \ldots, \theta_k) = \prod_{i=1}^{n} f_X(x_i) \qquad \text{if the population is continuous}$$

$$= \prod_{i=1}^{n} p_X(x_i) \qquad \text{if the population is discrete,}$$

where in either case $\Pi f_X(x_i)$ or $\Pi p_X(x_i)$ are evaluated at the observed sample values, $x_1, x_2, \ldots, x_n$.

Notice then that the likelihood function is a function only of the unknown parameters. If we let Ω be the set of possible values for $\theta_1, \theta_2, \ldots, \theta_k$, then the domain of L is Ω. If the population is discrete, L provides the probability of observing the actual sample values, as a function of $\omega \in \Omega$. If the population is continuous, L assigns the value of the joint density function of the sample values to each $\omega \in \Omega$, and the value of this joint density, of course, is proportional to the probability of observing sample values in a neighborhood of those that actually occurred. Thus, in either case, we might think of L as providing the probability of the sample occurring as a function of $\omega = (\theta_1, \theta_2, \ldots, \theta_k) \in \Omega$. If we can find that particular $\hat{\omega} = (\hat{\theta}_1, \hat{\theta}_2, \ldots, \hat{\theta}_k) \in \Omega$ which maximizes L, then $\hat{\theta}_1, \hat{\theta}_2, \ldots, \hat{\theta}_k$ would seem logical candidates as estimates for $\theta_1, \theta_2, \ldots, \theta_k$, respectively. This is precisely what the method of maximum likelihood accomplishes, as we can see in the following definition.

DEFINITION 3.2.2. The values of $\theta_1, \theta_2, \ldots, \theta_k$ which jointly maximize $L(\theta_1, \theta_2, \ldots, \theta_k)$ for $(\theta_1, \theta_2, \ldots, \theta_k) \in \Omega$, are the *maximum likelihood estimates* of the parameters. Since the likelihood function has been evaluated at the sample values, $x_1, x_2, \ldots, x_n$, generally the maximizing values of $\theta_1, \theta_2, \ldots, \theta_k$ will be functions of $x_1, x_2, \ldots, x_n$, say

$$\hat{\theta}_i = m_i(x_1, x_2, \ldots, x_n), \qquad i = 1, 2, \ldots, k.$$

Then the *maximum likelihood estimators* are the same functions of the elements of the sample; i.e., $\hat{\Theta}_i = m_i(X_1, X_2, \ldots, X_n)$. These quantities again are random variables whose behavior can be examined from sample to sample.

We shall now examine the same two examples mentioned earlier to see what the maximum likelihood estimators are for the same parameters. If we can find the values of $\theta_1, \theta_2, \ldots, \theta_k$ that maximize $K = \log L$, these are of course the same values that maximize L, since the log function is monotonic increasing. In many cases it is easier to maximize K.

Example 3.2.3

In the case of sampling from a $N(\mu, \sigma)$ population, the likelihood function is

$$L(\mu, \sigma^2) = \prod_{i=1}^{n} f_X(x_i) = \left(\frac{1}{2\pi\sigma^2}\right)^{n/2} e^{-(1/2\sigma^2)\sum_{i=1}^{n}(x_i-\mu)^2}, \qquad \mu \text{ real}, \qquad \sigma > 0.$$

Then

$$K(\mu, \sigma^2) = \log L(\mu, \sigma^2) = -\frac{n}{2}\log 2\pi - \frac{n}{2}\log \sigma^2 - \frac{1}{2\sigma^2}\sum_{i=1}^{n}(x_i - \mu)^2.$$

Since K is obviously a continuous function of μ and σ^2, we might expect to find its maximum value by solving the equations $\partial K/\partial\mu = 0$ and $\partial K/\partial\sigma^2 = 0$.

We find

$$\frac{\partial K}{\partial \mu} = \frac{1}{\sigma^2}\sum_{i=1}^{n}(x_i - \mu)$$

$$\frac{\partial K}{\partial \sigma^2} = -\frac{n}{2\sigma^2} + \frac{1}{2(\sigma^2)^2}\sum_{i=1}^{n}(x_i - \mu)^2.$$

Setting these equal to zero and solving simultaneously yields $\hat{\mu} = \bar{x}$, $\hat{\sigma}^2 = (1/n)\sum_{i=1}^{n}(x_i - \bar{x})^2$, the same as the method of moments estimates. Notice that $\partial^2 K/\partial \mu^2 = -n/\sigma^2$

$$\frac{\partial^2 K}{\partial \mu\, \partial \sigma^2} = -\frac{1}{(\sigma^2)^2}\sum(x_i - \mu)$$

$$\frac{\partial^2 K}{\partial (\sigma^2)^2} = \frac{n}{2(\sigma^2)^2} - \frac{1}{(\sigma^2)^3}\sum(x_i - \mu)^2$$

and that the 2×2 matrix

$$A = \begin{bmatrix} \dfrac{\partial^2 K}{\partial \mu^2} & \dfrac{\partial^2 K}{\partial \mu\, \partial \sigma^2} \\ \dfrac{\partial^2 K}{\partial \mu\, \partial \sigma^2} & \dfrac{\partial^2 K}{(\partial \sigma^2)^2} \end{bmatrix}$$

evaluated at $\hat{\mu}$, $\hat{\sigma}^2$ is negative definite. Thus $\hat{\mu}$, $\hat{\sigma}^2$ are the coordinates which maximize K (and L). The estimators for μ and σ^2 then are $\bar{X}$ and

$$[(n - 1)/n]S^2,$$

respectively.

Example 3.2.4

Assume again that we have a sample of a uniform random variable from the interval (a, b). Then the likelihood function for the sample is

$$L(a, b) = \prod_{i=1}^{n} f_X(x_i) = \frac{1}{(b - a)^n}, \qquad a \le x_i \le b, \qquad i = 1, 2, \ldots, n.$$

In particular, then, $a \le x_{(1)}$, where $x_{(1)}$ is the smallest sample value and $b \ge x_{(n)}$, where $x_{(n)}$ is the largest sample value; thus $L(a, b) = 1/(b - a)^n$, $a \le x_{(1)}$, $b \ge x_{(n)}$. Clearly, L is a decreasing function of $b - a$; thus, to maximize L we would like $b - a$ to be as small as possible. Since we require $a \le x_{(1)}$, $b \ge x_{(n)}$, clearly $\hat{b} - \hat{a}$ is as small as possible with $\hat{b} = x_{(n)}$, $\hat{a} = x_{(1)}$. Thus the maximum likelihood estimators for a and b are $\hat{A} = X_{(1)} = \min(X_1, X_2, \ldots, X_n)$, $\hat{B} = X_{(n)} = \max(X_1, X_2, \ldots, X_n)$. Notice that these estimators cannot have the defect mentioned for the method of moments estimators in Example 3.2.2.

The following theorem gives a frequently used property of maximum likelihood estimators.

THEOREM 3.2.1. Assume that a population distribution depends on k unknown parameters, $\theta_1, \theta_2, \ldots, \theta_k$ and let $\lambda_1, \lambda_2, \ldots, \lambda_r$, $r \leq k$, be r new parameters, each of which is a function of $\theta_1, \theta_2, \ldots, \theta_k$; that is $\lambda_i = h_i(\theta_1, \theta_2, \ldots, \theta_k)$, $i = 1, 2, \ldots, r$. Then if $\hat{\theta}_1, \hat{\theta}_2, \ldots, \hat{\theta}_k$ are the maximum likelihood estimates of $\theta_1, \theta_2, \ldots, \theta_k$, $\hat{\lambda}_i = h_i(\hat{\theta}_1, \hat{\theta}_2, \ldots, \hat{\theta}_k)$, $i = 1, 2, \ldots, r$, are the maximum likelihood estimates of $\lambda_1, \lambda_2, \ldots, \lambda_r$.

Proof: Notice that in defining $\lambda_1, \ldots, \lambda_r$ as functions of $\theta_1, \theta_2, \ldots, \theta_k$, we have defined a new parameter space Ω_λ of values for the λ_i's, which is a transformation of the original space Ω_θ for the θ_j's. Since L is a function of the elements of Ω_θ, we can define

$$M(\lambda_1, \lambda_2, \ldots, \lambda_r) = \max_{\lambda_i = h_i(\theta_1, \ldots, \theta_k)} L(\theta_1, \theta_2, \ldots, \theta_k)$$

to be the (natural) likelihood function for $\omega \in \Omega_\lambda$. Note that we may have more than one set of values for $\theta_1, \theta_2, \ldots, \theta_k$ corresponding to the *same* point $\lambda_1, \lambda_2, \ldots, \lambda_r$; hence, we would have several values of L for this point and, in such a case, we let M equal the biggest of those values. If the transformation from Ω_θ to Ω_λ is 1 to 1, hence $r = k$, then there is only one value of L for each point in Ω_λ and we set M equal to that value. It then follows trivially that if $\hat{\theta}_1, \hat{\theta}_2, \ldots, \hat{\theta}_k$ maximize L, then

$$\hat{\lambda}_i = h_i(\hat{\theta}_1, \hat{\theta}_2, \ldots, \hat{\theta}_k), \qquad i = 1, 2, \ldots, r$$

maximize M and the result is established.

This result is useful in many different instances in applied statistics. For example, in the case of a normal population, the maximum likelihood estimator of σ then is $S\sqrt{(n-1)/n}$. The following example of *probit analysis* presents a less trivial use of this result.

Example 3.2.5 (Probit Analysis)
The estimation of the tolerance of a particular type of biological organism to a specific toxin presents an interesting problem. For example, suppose we are given a population of a certain type of fly and a chemical agent which is toxic to this type of fly if applied in large enough concentrations. Since the concentration used can be varied essentially continuously, it is then of interest to know what proportion of the population will be killed as the concentration is varied. As the concentration approaches zero, certainly

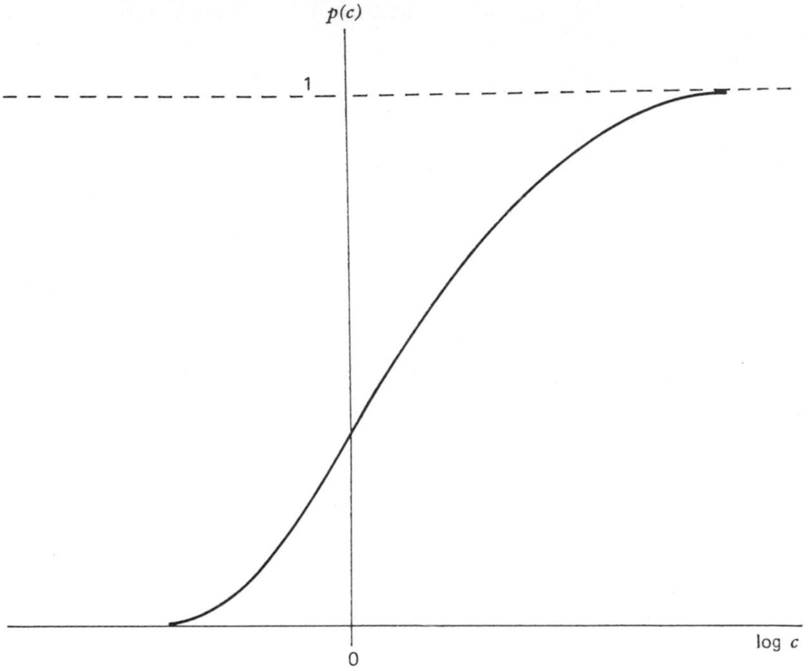

FIGURE 3.2.1. Assumed relation between proportion killed and concentration *c*.

the proportion killed should also approach zero. Thus, the concentration c can be varied at will, and we ask what is the proportion, $p(c)$, of the population which will be killed at this concentration.

The standard probit analysis model assumes that $p(c)$ varies with $\log c$ like a normal distribution function (see Figure 3.2.1). That is, there is assumed to be an underlying normal random variable W such that

$$P(W \leq \log c) = p(c).$$

This assumption corresponds to an increase in concentration from c_1 to $c_2 > c_1$ causing an increase in the proportion killed being equal to an area under a normal curve between $\log c_1$ and $\log c_2$. See Figure 3.2.2. Of course, the location of this normal density (or distribution) function is not known, nor is its variance; indeed, these quantities can be varied in any desired way by changing the units used in measuring concentration.

Now, since

$$P(W \leq \log c) = P\left(Z \leq \frac{\log c - \mu_W}{\sigma_W}\right) = p(c)$$

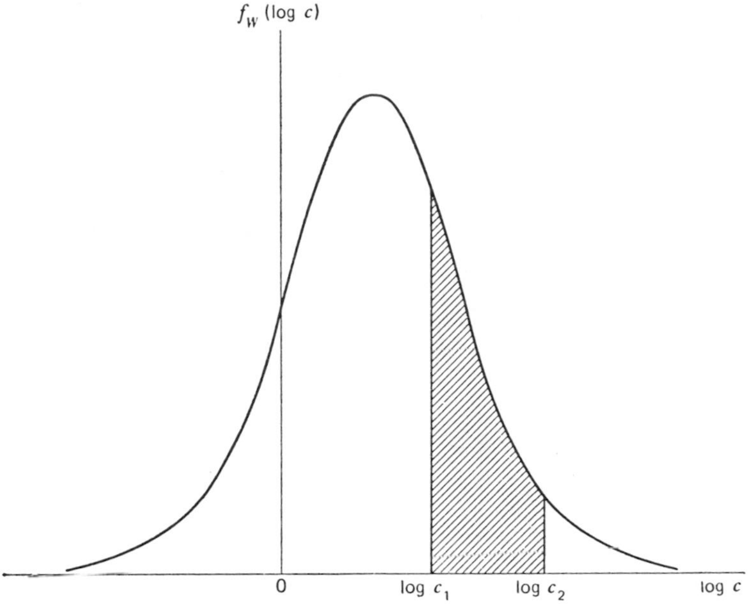

FIGURE 3.2.2. Normal density of tolerance in the population.

where Z is $N(0, 1)$, μ_W and σ_W are the parameters of the distribution of W, note that from the probability integral transform (see Exercise 1.3.15), we have

$$N_Z^{-1}(p(c)) = \frac{\log c - \mu_W}{\sigma_W}$$
$$= \alpha + \beta \log c,$$

where N_Z is the $N(0, 1)$ distribution function, $\alpha = -(\mu_W/\sigma_W)$, $\beta = 1/\sigma_W$. That is, the relationship between the inverse standard normal distribution function at $p(c)$ and $\log c$ is a straight line (see Figure 3.2.3). Given the type of insect and the specific toxin, it is then of interest to estimate the parameters of this straight line, α and β. The usual method of doing this is to apply maximum likelihood.

Assume that k different concentrations, $c_1, c_2, \ldots, c_k$ are used; concentration c_i is applied to a batch of n_i insects of the given type, $i = 1, 2, \ldots, k$. Let $p_i = N_Z(\alpha + \beta c_i)$ be the true proportion of the population that would be killed with concentration c_i and let f_i be the observed proportion killed at concentration c_i. (Thus, if x_i is the number killed, $f_i = x_i/n_i$, $x_i = 0, 1, 2, \ldots, n_i$.) It is clear that $n_i f_i = x_i$ is binomial with parameters n_i and p_i,

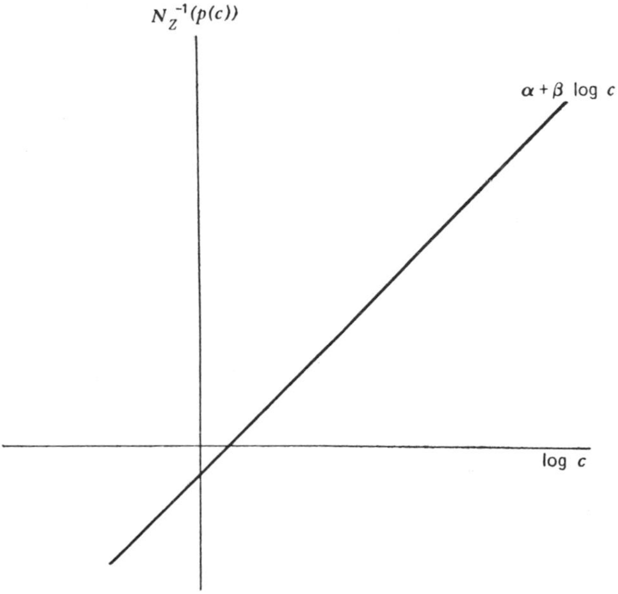

FIGURE 3.2.3. Relationship between $N_Z^{-1}(p(c))$ and log *c*.

$i = 1, 2, \ldots, k$, if we assume that each of the individual flies is either killed or not, independently, at the given dosage. Furthermore, it is reasonable that the observations from batch to batch are independent, if they are handled separately. Thus, the likelihood function of the sample is

$$L(\alpha, \beta) = \prod_{i=1}^{k} \binom{n_i}{n_i f_i} p_i^{n_i f_i}(1 - p_i)^{n_i(1-f_i)}$$

and the log of the likelihood is

$$K = \log L = \sum_{i=1}^{k} \left\{ \log \binom{n_i}{n_i f_i} + n_i f_i \log p_i + n_i(1 - f_i)\log(1 - p_i) \right\}$$

By the chain rule for partial derivatives

$$\frac{\partial K}{\partial \alpha} = \sum_{i=1}^{k} \frac{\partial K}{\partial p_i} \frac{\partial p_i}{\partial \alpha}, \qquad \frac{\partial K}{\partial \beta} = \sum_{i=1}^{k} \frac{\partial K}{\partial p_i} \frac{\partial p_i}{\partial \beta};$$

also, since $p_i = N_Z(\alpha + \beta \log c_i)$,

$$\frac{\partial p_i}{\partial \alpha} = n_Z(\alpha + \beta \log c_i), \qquad \frac{\partial p_i}{\partial \beta} = n_Z(\alpha + \beta \log c_i)\log c_i,$$

where n_Z is the standard normal density function, and

$$\partial K/\partial p_i = \frac{n_i(f_i - p_i)}{p_i(1 - p_i)}.$$

Thus, the equations to be solved for the estimates of α and β are

$$\sum_{i=1}^{k} \frac{n_i(f_i - p_i)}{p_i(1 - p_i)} n_Z(\alpha + \beta \log c_i) = 0$$

$$\sum_{i=1}^{k} \frac{n_i(f_i - p_i)}{p_i(1 - p_i)} \log c_i n_Z(\alpha + \beta \log c_i) = 0$$

where $p_i = N_Z(\alpha + \beta \log c_i)$. Since these equations are far from linear, it is not a simple matter to determine their solutions. Generally, the points $(\log c_i, N_Z^{-1}(f_i))$ are plotted, a straight line is drawn in by hand to determine initial values, α_1 and β_1, for α and β, and then an iterative technique is applied which converges to the values $\hat{\alpha}$ and $\hat{\beta}$ at which the above two equations are satisfied (see Exercise 3.2.21 below). Once the values of $\hat{\alpha}$ and $\hat{\beta}$ have been determined, then the maximum likelihood estimates for μ_W and σ_W are $\hat{\mu}_W = -\hat{\alpha}/\hat{\beta}$, $\hat{\sigma}_W = 1/\hat{\beta}$ by the invariance of maximum likelihood (Theorem 3.2.1.).

Example 3.2.6 (Reliability Analysis)
For many component parts used in complex pieces of equipment, the exponential distribution has historically proved quite satisfactory as an assumed model for the time to failure. Thus, for a given transistor, for example, we might assume that its time to failure is an exponential random variable with parameter λ, where $\lambda > 0$ is, of course, unknown. To estimate λ we could put n of these transistors on test independently (realistically representing the conditions of the intended usage), wait until all n had failed, and then use this sample of n observations of the underlying exponential random variable to estimate λ with maximum likelihood (see Exercise 3.2.6 below for the form of this estimator). Especially if λ is fairly small, however, this sort of approach may take a considerable amount of time; it seems wasteful to not be able to estimate λ if $n - 1$ of the transistors have failed after 100 hours, say, but we have to wait for that last one to complete our estimation.

There are many alternative ways of conducting the testing, rather than waiting until all n failures have occurred. We might, for example, put n items on test and conclude the test when the first item fails. Now, if the items have been tagged 1 through n, initially, it is plain that we have observed the value of one of the random variables X_1 through X_n when the first item fails and we only have partial information on the other $(n - 1)$. (We know their times to failure exceed the observed minimum time to failure.) Remembering that the likelihood function is essentially proportional to the probability of occurrence of the sample, we see that it equals

$$L(\lambda) = \binom{n}{1} \lambda e^{-\lambda x}[1 - F_X(x)]^{n-1} = n\lambda e^{-n\lambda x},$$

where x is the observed time to failure of the first item. The combinatorial coefficient $\binom{n}{1}$ occurs in L because any one of the n items could be the first to fail, $\lambda e^{-\lambda x}$ is the density function for the one observed failure time and $[1 - F_X(x)]^{n-1}$ is the probability that each of the remaining $n - 1$ items fail (independently) after x hours have passed. We find then that $\hat{\lambda} = 1/nx$ maximizes this likelihood function and is the maximum likelihood estimate for this method of conducting the test.

Another alternative way of testing items is, instead of waiting until some specified number of items have failed, to simply have some preplanned length of time T_0 at which the test will be concluded regardless of how many have failed by that time (unless, of course, all have already failed). If we assume that exactly k of the items have failed by time T_0, say at $x_1 \leq x_2 \leq \cdots \leq x_k \leq T_0$ hours, $k = 0, 1, 2, \ldots, n - 1$, then the likelihood of the sample is

$$L(\lambda) = \binom{n}{k} \lambda^k e^{-\lambda \sum_{i=1}^{k} x_i}[1 - F_X(T_0)]^{n-k}$$

$$= \binom{n}{k} \lambda^k e^{-\lambda[\sum x_i + (n-k)T_0]}.$$

We find the maximum of this function to be at

$$\hat{\lambda} = \frac{k}{\sum_{i=1}^{k} x_i + (n - k)T_0},$$

which then is the maximum likelihood estimate for λ.

This example illustrates a very important fact for applied statistics. The method in which the testing is accomplished, or the sample is taken, has an effect on the likelihood function and thus has an effect on the estimate (or other statistical technique) which is appropriate. In many instances it is not as easy to see what the specific effect is or should be, but nonetheless we should be aware of its presence.

EXERCISE 3.2

In each of Exercises 1 through 12, determine (a) the method of moments estimator(s), and (b) the maximum likelihood estimator(s) of the parameters of the given distribution, based on a random sample of n observations.

1. X is Bernoulli with parameter p.

2. X is Poisson with parameter λ.

3. X is uniform on $(b - 1/2, b + 1/2)$.

4. X is geometric with parameter p.

5. X is negative binomial with parameters r (known) and p (unknown).

6. X is exponential with parameter λ.

7. X is gamma with parameters r (known) and λ (unknown).

8. X is binomial with parameters N (known) and p (unknown).

9. X is bivariate normal with parameters $\mu_1, \mu_2, \sigma_{11}, \sigma_{12}, \sigma_{22}$.

10. X is bivariate normal with parameters $\mu_1, \mu_2, \sigma_1^2, \sigma_2^2, \rho$.

11. X is normal with mean μ and variance μ^2.

12. X is normal with mean μ and variance μ.

13. For each of Exercises 1 through 8, what is the maximum likelihood estimator of the mean and variance of X?

14. A used-car salesman assumes the number of sales he makes per day is a Poisson random variable. His maximum likelihood estimate of λ, from his past year, is $\hat{\lambda} = 1/3$ (cars per day). What is the maximum likelihood estimate of the probability he will sell at least one car today, given the previous information?

15. The time to failure of a certain electron tube is an exponential random variable. From a previous sample, the maximum likelihood estimate of λ was $\hat{\lambda} = .005$. What is the maximum likelihood estimate of the median time to failure for this type of tube?

16. Show that if $n = 2$, for the bivariate normal random variable given in Exercise 3.2.10 above, then $P(\hat{\rho} = -1) + P(\hat{\rho} = 1) = 1$ and, if $\rho = 0$, each of these probabilities is 1/2.

17. $\mathbf{X}$ is a multinomial random vector with parameters $n, p_1, p_2, \ldots, p_k$; given that in a certain experiment $\mathbf{X} = \mathbf{x}$, what are the estimators of $p_1, p_2, \ldots, p_k$, (a) using the method of moments and (b) using maximum likelihood?

18. In problem 17, assume $k = 3$ and that $p_1 = 2\theta, p_2 = 3\theta, p_3 = 1 - 5\theta$. What is the maximum likelihood estimator for θ?

19. Assume that weights of adults in the United States form a normal population and that a sample yielded 170 pounds and 25 pounds, respectively, as the maximum likelihood estimates for μ and σ. What is the maximum likelihood estimate of the weight exceeded by only 10% of the population? What is the maximum likelihood estimate of the proportion of the population with weights between 140 and 190 pounds?

20. In the example about *probit analysis* (3.2.5) assume $k = 2$ different concentrations are used and apply the method of moments to estimate α and β. What difficulty might you encounter if $k > 2$?

21. The usual iterative technique used to solve the likelihood equations which arise in probit analysis is the following: The equations to be solved are

$$\sum_{i=1}^{k} \frac{n_i(f_i - p_i)}{p_i(1 - p_i)} \frac{\partial p_i}{\partial \alpha} = 0$$

$$\sum_{i=1}^{k} \frac{n_i(f_i - p_i)}{p_i(1 - p_i)} \frac{\partial p_i}{\partial \beta} = 0,$$

where $\partial p_i/\partial \alpha = g(\alpha, \beta)$, $\partial p_i/\partial \beta = h(\alpha, \beta)$. Using the Taylor series expansion for functions of two variables we can write

$$g(\alpha, \beta) \doteq g(\alpha_1, \beta_1) + (\alpha - \alpha_1)\frac{\partial^2 p_i}{\partial \alpha^2} + (\beta - \beta_1)\frac{\partial^2 p_i}{\partial \alpha\, \partial \beta}$$

and we have an analogous equation for $h(\alpha, \beta)$; the second pure and mixed partial derivatives and the p_i are also evaluated at α_1, β_1, the initial guess arrived at graphically. Then, replacing $\partial p_i/\partial \alpha$ and $\partial p_i/\partial \beta$ by these Taylor series approximations, notice that we then have two linear equations in the two unknowns α and β which can easily be solved for the second order estimates α_2, β_2. Then, these quantities are used in place of α_1 and β_1, as above, to get the third order solutions, etc. Once two successive sets of solutions are essentially equal, the procedure is terminated, having converged to the estimates $\hat{\alpha}$ and $\hat{\beta}$ which satisfy the likelihood equations. Given the following set of data, apply this procedure to estimate α and β:

$\log c_i$	n_i	x_i	f_i
−1	20	2	.1
0	25	10	.4
1	10	8	.8

22. In the probit analysis example, show that if $k = 2$, then the maximum likelihood estimates of α and β are not given by the line which passes through the two observed points $(\log c_1, N_Z^{-1}(f_1))$, $(\log c_2, N_Z^{-1}(f_2))$.

23. In example 3.2.6, show that if the n items are to be tested T_o hours, and exactly 1 failure occurs at $x = T_o$ hours, then the two estimates given in that example are identical.

24. Show that in the second part of Example 3.2.6 (planned test time of T_o hours) there is no reason to not allow $k = n$; that is, compare the estimator from Exercise 3.2.6 with the estimator given in the example for $k = n$.

25. Generalize the first part of Exercise 3.2.6; that is, n items are put on test and once $r > 1$ items have failed the test is concluded. What is the maximum likelihood estimate for λ?

26. Assume the time to failure for a given type of item is uniform on the interval $(0, b)$. If items are tested in the manner described in Example 3.2.6, what are the maximum likelihood estimates for b?

27. Given X is uniform (a, b), show that $a = \mu_X - \sigma_X\sqrt{3}$, $b = \mu_X + \sigma_X\sqrt{3}$ and compare with Example 3.2.2.

28. Assume $X_1, X_2, \ldots, X_n$ is a random sample of a uniform random variable on $(-b, b)$ and find the method of moments estimator for b and the maximum likelihood estimator for b. This is one of those cases in which only one equation is needed to determine the method of moments estimate for b, but that one equation is not $E[X] = \bar{X}$.

29. Show that the matrix of second partials given in Example 3.2.3 is negative definite and thus the likelihood function does have a maximum at $\hat{\mu}$, $\hat{\sigma}^2$.

30. Suppose we have a sample of n_1 observations from a normal population with mean μ and variance σ_1^2 and a sample of n_2 observations from a second normal population with mean μ and variance σ_2^2. (Note that the two means are equal.) What are the maximum likelihood estimators for μ, σ_1^2, and σ_2^2?

31. Assume that a box contains $3N$ flies, N female of the same genotype, N male of genotype 1 and N male of genotype 2. Each female mates with one male, then is placed in a separate box so that all her descendants can be identified. Each descendant has characteristic A or B. If she mated with male type 1, genetic theory predicts that the probability a particular descendant has characteristic A is q_1; thus the probability a descendant has characteristic B is $1 - q_1$ (q_1 is known). Similarly, if she mated with a male type 2 the probability a particular descendant has characteristic A is q_2 (known), that it has characteristic B is $1 - q_2$. Assume that the ith female has n_i descendants, $i = 1, 2, \ldots, N$, each independently bearing characteristic A or B. Furthermore, assume that for each female there is a common probability p that she mated with a male type 1 (preference probability) and that p is unknown. Furthermore, the females choose their male mates independently. Given that of the n_i dependents of female i, X_i have characteristic A, $i = 1, 2, \ldots, N$, how would you determine the maximum likelihood estimator for p?

32. A university high jumper has a decreasing probability of successfully clearing the bar, as the bar is moved up. In fact, the probit analysis model, without taking logs, might be quite appropriate. Assume that a given jumper made 10 attempts, 8 of which were successful, at 6 feet and 10 attempts, 3 of which were successful, at 6 feet, 6 inches. Use the technique of problem 22 to estimate α and β, then μ

and σ. Do these estimates differ from the maximum likelihood estimates of μ and σ for this same case?

3.3 Properties of Estimators

We saw in the last section two different ways of deriving estimates of the values of unknown parameters; there are other methods that are quite appropriate as well in specific instances. (We shall study two more methods: Bayesian estimates and least squares estimates in succeeding chapters.) Given that in the same problem different methods of estimation can give rise to different estimates for the same parameter, it becomes important to derive methods of comparing estimators. Then, the estimator which has the better properties should logically be the one that would be used in a given problem. In this section we shall examine some of the common methodology which has been developed over the years for comparing estimators. It is well to keep in mind that we are studying properties of *estimators*, not estimates. For a given single sample it is quite possible that the estimator which does not have good properties in general, might take on the "best" value and thus give rise to the best estimate for that particular sample. But, since we do not know the actual values of the unknown parameters we of course are completely unable to judge the goodness of the estimates for a particular sample. We are, however, able to study the behavior of the estimators, the random variables, over repeated samples. This study of behavior over repeated samples is the way in which estimators are compared.

The first property we shall examine is that of unbiasedness, defined below.

DEFINITION 3.3.1. Θ is an *unbiased estimator* for θ if $E[\Theta] = \theta$. The difference $b_\Theta(\theta) = E[\Theta] - \theta$ is called the *bias* of Θ.

Thus, if the mean value of the distribution for Θ is θ, we say that Θ is an unbiased estimator for θ; this can also be interpreted as saying that over repeated samples, the long run average value of Θ is θ, the parameter estimated. For any particular sample the value of Θ, the estimate, may or may not be close to the true unknown value θ. As we saw in Section 3.2, both the method of moments and maximum likelihood yield $\bar{X}$ and $[(n-1)/n]S^2$, respectively, as the estimators for μ and σ^2 in a normal population. Since $E[\bar{X}] = \mu$ and

$$E\left[\frac{n-1}{n} S^2\right] = \sigma^2 - (\sigma^2/n),$$

note that either of these methods may give rise to unbiased estimators, but that they do not necessarily do so. Since $\lim_{n\to\infty} [(n-1)/n]\sigma^2 = \sigma^2$, we say that $[(n-1)/n]S^2$ is an *asymptotically unbiased estimator.*

Certainly other aspects of the distribution of an estimator, besides just its mean value, are worthy of study for comparing estimators. A consistent estimator is defined below.

DEFINITION 3.3.2. Θ is a *consistent estimator* for θ if its distribution converges in probability to θ. That is, if for any $\varepsilon > 0$,

$$\lim_{n\to\infty} P(|\Theta - \theta| > \varepsilon) = 0,$$

where n is the sample size.

Rather than using this definition directly, the following theorem gives a relatively easy method for checking the consistency of an estimator.

THEOREM 3.3.1. Θ is a consistent estimator of θ if (*i*) $\lim_{n\to\infty} E[\Theta] = \theta$ and (*ii*) $\lim_{n\to\infty} \sigma_\Theta^2 = 0$, where n is the sample size.

Proof: Suppose $\lim_{n\to\infty} E[\Theta] = \theta$ and $\lim_{n\to\infty} \sigma_\Theta^2 = 0$. Then $\lim_{n\to\infty} b_\Theta(\theta) = 0$, where $b_\Theta(\theta)$ is the bias in Θ. From Chebycheff's inequality, for any n,

$$P[|\Theta - E(\Theta)| < k\sigma_\Theta] \geq 1 - \frac{1}{k^2}.$$

Furthermore, since $\Theta - E[\Theta] = \Theta - \theta - b_\Theta(\theta)$

$$P[|\Theta - \theta| < k\sigma_\Theta + |b_\Theta(\theta)|] \geq P[|\Theta - E[\Theta]| < k\sigma_\Theta].$$

If we now choose $k = (\varepsilon - |b_\Theta(\theta)|)/\sigma_\Theta$, for any fixed ε, note that

$$P[|\Theta - \theta| < \varepsilon] \geq 1 - \frac{\sigma_\Theta^2}{[\varepsilon - b_\Theta(\theta)]^2}$$

and

$$\lim_{n\to\infty} P[|\Theta - \theta| < \varepsilon] \geq \lim_{n\to\infty}\left[1 - \frac{\sigma_\Theta^2}{[\varepsilon - b_\Theta(\theta)]^2}\right] = 1;$$

thus Θ does converge in probability to θ.

Example 3.3.1

Given a random sample from a normal population we have seen that $\bar{X}$ has mean μ and variance σ^2/n and that S^2 has mean σ^2 and variance

$$2\sigma^4/(n-1);$$

thus $[(n-1)/n]S^2$ has mean $[(n-1)/n]\sigma^2$ and variance $[2(n-1)/n^2]\sigma^4$. Note then that $[(n-1)/n]S^2$ is a consistent estimator for σ^2 (as is S^2), since it is asymptotically unbiased and its variance goes to 0. $\bar{X}$ is a consistent estimator for μ.

Consistency is itself an asymptotic property, making statements about an estimator in the limit as the sample size increases indefinitely. For large samples it may be of great interest to ensure that the estimator used is consistent, to hopefully avoid using an estimate that may completely miss the mark. For small samples, though, the property of consistency may not be particularly relevant.

Estimators based on sufficient statistics have desirable properties regardless of the sample size. The definition of a sufficient statistic is given below.

DEFINITION 3.3.3. Given a random sample $X_1, X_2, \ldots, X_n$ from a population whose distribution depends on a parameter θ, the statistic

$$Y_1 = g_1(X_1, X_2, \ldots, X_n)$$

is *sufficient* for θ if and only if, for any other set of statistics $Y_j = g_j(X_1, X_2, \ldots, X_n)$, $j = 2, 3, \ldots, n$, such that the transformation from the X_k's to the Y_i's has a nonzero jacobian, the conditional distribution of $Y_2, Y_3, \ldots, Y_n$, given $Y_1 = y_1$, does not depend on θ.

As we know, given the original random variables $X_1, X_2, \ldots, X_n$, we can transform their distribution to that of the statistics $Y_1, Y_2, \ldots, Y_n$ in an unlimited number of ways, provided only that the jacobian of the transformation is nonzero. Thus, if we think of Y_1 as being a fixed statistic and the remaining elements of the transformation $Y_2, Y_3, \ldots, Y_n$ as being free to vary, then Y_1 is sufficient if the resulting conditional distribution for Y_2 through Y_n, given $Y_1 = y_1$, does not depend on θ. Since this conditional distribution does not depend on θ, it is plain that no aspect of the distributions of Y_2 through Y_n could depend on θ; thus, $Y_2, Y_3, \ldots, Y_n$ do not contain any additional information about θ, beyond that given by $Y_1 = y_1$. In a sense, then, Y_1 contains all the information in the sample about θ and is thus called the *sufficient statistic* for θ. The following example gives a simple instance in which we have a sufficient statistic for the parameter p of a Bernoulli random variable.

Example 3.3.2

Assume X_1, X_2 are independent Bernoulli random variables, each with parameter p. Then

$$p_{X_1,X_2}(x_1, x_2) = p^{x_1+x_2}(1-p)^{2-x_1-x_2}, \qquad x_1 = 0, 1, \qquad x_2 = 0, 1.$$

Let $y_1 = x_1 + x_2$, $y_2 = g_2(x_1, x_2)$, where $\partial g_2/\partial x_1 \neq \partial g_2/\partial x_2$, so that the jacobian of the transformation from (x_1, x_2) to (y_1, y_2) is not zero. Then, the joint probability function for $Y_1 = X_1 + X_2$, $Y_2 = g_2(X_1, X_2)$ is positive only at the points $(0, g_2(0, 0))$, $(1, g_2(1, 0))$, $(1, g_2(0, 1))$ and $(2, g_2(1, 1))$ and has values $(1-p)^2$, $p(1-p)$, $p(1-p)$, p^2 at these points, respectively.

The marginal probability function for Y_1 is

$$p_{Y_1}(0) = (1 - p)^2$$
$$p_{Y_1}(1) = 2p(1 - p)$$
$$p_{Y_1}(2) = p^2$$

and 0 everywhere else. Then, the conditional probability function for $Y_2 = g_2(X_1, X_2)$, given $Y_1 = 0$, is

$$p_{Y_2|Y_1}(g_2(0, 0) \mid Y_1 = 0) = \frac{p_{Y_1,Y_2}(0, g_2(0, 0))}{p_{Y_1}(0)} = \frac{(1 - p)^2}{(1 - p)^2} = 1$$

and 0, otherwise; similarly, given $Y_1 = 1$,

$$p_{Y_2|Y_1}(g_2(1, 0) \mid Y_1 = 1) = p_{Y_2|Y_1}(g_2(0, 1) \mid Y_1 = 1) = \frac{p(1 - p)}{2p(1 - p)} = \frac{1}{2}$$

and

$$p_{Y_2|Y_1}(g_2(1, 1) \mid Y_1 = 2) = \frac{p^2}{p^2} = 1.$$

Notice that, no matter which value we are given for Y_1, and no matter which $Y_2 = g_2(X_1, X_2)$ we use, the conditional distribution for Y_2 given Y_1 does not involve p. Thus, having taken the sample and observed the value of Y_1, there is no reason to consider any second statistic (function of the sample values); its (conditional) distribution does not involve p and thus cannot provide any additional information for making inferences about p, given $Y_1 = y_1$. Y_1 is a sufficient statistic for p. In the exercises below, you are asked to explain the reason for the requirement that the jacobian of the transformation should not be zero.

As might be apparent from this example, applying the definition of a sufficient statistic directly may not be an easy task, especially for arbitrary sample sizes. The following theorem provides a procedure which frequently makes it a fairly simple task to investigate the sufficiency of a statistic. (This is called the Fisher-Neyman *factorization criterion.*) We shall examine this theorem only for continuous random variables; the proof in the discrete case is similar.

THEOREM 3.3.2. Assume $X_1, X_2, \ldots, X_n$ is a random sample of a continuous random variable X which has a single unknown parameter θ. Let $Y_1 = g_1(X_1, X_2, \ldots, X_n)$ be a statistic. Then Y_1 is a sufficient statistic for θ if

$$\prod_{i=1}^{n} f_X(x_i) = G(g_1(x_1, \ldots, x_n))H(x_1, x_2, \ldots, x_n),$$

where H does not depend on θ in any way. (G of course does involve θ.)

Proof: Suppose, as above, that

$$\prod_{i=1}^{n} f_X(x_i) = G(g_1(x_1, \ldots, x_n))H(x_1, x_2, \ldots, x_n).$$

Then, let $y_i = g_i(x_1, \ldots, x_n)$, $i = 2, \ldots, n$, be any additional variables (not involving θ) such that the transformation from $x_1, x_2, \ldots, x_n$ to $y_1, y_2, \ldots, y_n$ has jacobian $|J| \neq 0$. Then the joint density for $Y_i = g_i(X_1, X_2, \ldots, X_n)$, $i = 1, 2, \ldots, n$, is $\prod_{i=1}^{n} f_X(x_i)\,|J|$, where the x_i's have been replaced by their equivalents in $y_1, y_2, \ldots, y_n$. But this is

$$G(y_1)H(x_1, \ldots, x_n)\,|J|$$

where again the x_i's in H are expressed in terms of $y_1, y_2, \ldots, y_n$. Then the marginal density for y_1 is

$$\begin{aligned} f_{Y_1}(y_1) &= \int \cdots \int_R G(y_1)H(x_1, \ldots, x_n)\,|J| \prod_{j=2}^{n} dy_j \\ &= G(y_1) \int \cdots \int_R H(x_1, \ldots, x_n)\,|J| \prod_{j=2}^{n} dy_j \end{aligned}$$

where R is the range of $y_2, y_3, \ldots, y_n$, for the given value of y_1. The conditional density for $Y_2, \ldots, Y_n$, given $Y_1 = y_1$, then is

$$\begin{aligned} f_{Y_2,\ldots,Y_n}(y_2, y_3, \ldots, y_n \mid y_1) &= \frac{G(y_1)H(x_1, \ldots, x_n)\,|J|}{G(y_1) \int \cdots \int_R H(x_1, \ldots, x_n)\,|J| \prod dy_j} \\ &= \frac{H(x_1, x_2, \ldots, x_n)\,|J|}{\int \cdots \int_R H(x_1, \ldots, x_n)\,|J| \prod dy_j} \end{aligned}$$

which does not depend on θ in any way. Thus $Y_1 = g_1(X_1, \ldots, X_n)$ is a sufficient statistic for θ.

This theorem enables us to easily search for sufficient statistics for unknown parameters.

Example 3.3.3

In the Bernoulli sample of size 2 treated in Example 3.3.2, the density of the sample is

$$p^{x_1+x_2}(1-p)^{2-x_1-x_2} = p^y(1-p)^{2-y} \cdot 1 = G(y)H(x_1, x_2)$$

where $y = x_1 + x_2$, $H(x_1, x_2) = 1$ for all (x_1, x_2). Thus $Y = X_1 + X_2$ is a sufficient statistic for p.

Example 3.3.4

Assume $X_1, X_2, \ldots, X_n$ is a random sample of a normal random variable with mean μ and known variance σ^2. Then

$$\prod_{i=1}^{n} f_X(x_i) = \left(\frac{1}{2\pi\sigma^2}\right)^{n/2} e^{-\frac{1}{2}\sum_{i=1}^{n}[(x_i-\mu)^2/\sigma^2]}.$$

Now, $\sum_{i=1}^{n} (x_i - \mu)^2 = \sum_{i=1}^{n} (x_i - \bar{x})^2 + n(\bar{x} - \mu)^2$ and thus the joint density can be written

$$\begin{aligned}\prod_{i=1}^{n} f_X(x_i) &= e^{-[n(\bar{x}-\mu)^2/2\sigma^2]}\left(\frac{1}{2\pi\sigma^2}\right)^{n/2} e^{-\frac{1}{2}\sum(x_i-\bar{x})^2/\sigma^2} \\ &= G(\bar{x})H(x_1, x_2, \ldots, x_n)\end{aligned}$$

where H does not involve μ. Thus, for any given value of σ^2, $\bar{X}$ is a sufficient statistic for μ.

Granted that a sufficient statistic contains all the information in the sample about a parameter, it then would seem reasonable that we should use an estimator for the parameter which is a function of the sufficient statistic. The Rao-Blackwell theorem, given below, formalizes a good reason for doing this.

THEOREM 3.3.3. Let $X_1, \ldots, X_n$ be a random sample of a random variable with density which depends on θ. Let Θ be any unbiased estimator for θ and let Y_1 be the sufficient statistic for θ. Let $E[\Theta \mid Y_1 = y_1] = h(y_1)$. Then $E[h(Y_1)] = \theta$ and $\sigma^2_{h(Y_1)} \leq \sigma^2_{\Theta}$. That is, $h(Y_1)$ then is also an unbiased estimator for θ and its variance is no larger than that of Θ.

Proof:

$$\begin{aligned}E[\Theta \mid Y_1 = y_1] &= \int_{-\infty}^{\infty} u f_{\Theta|Y_1}(u \mid y_1)\, du \\ &= \int_{-\infty}^{\infty} u \frac{f_{\Theta,Y_1}(u, y_1)}{f_{Y_1}(y_1)}\, du = h(y_1).\end{aligned}$$

Thus

$$\int_{-\infty}^{\infty} u f_{\Theta,Y_1}(u, y_1)\, du = h(y_1) f_{Y_1}(y_1)$$

and

$$\begin{aligned}E[h(Y_1)] &= \int_{-\infty}^{\infty} \{h(y_1) f_{Y_1}(y_1)\}\, dy_1 \\ &= \int_{-\infty}^{\infty} \left\{\int_{-\infty}^{\infty} u f_{\Theta,Y_1}(u, y_1)\, du\right\} dy_1 = E[\Theta] = \theta,\end{aligned}$$

so $h(Y_1)$ is also an unbiased estimator for θ. By definition

$$\begin{aligned}\sigma_\Theta^2 = E[(\Theta - \theta)^2] &= E[(\{\Theta - h(Y_1)\} + \{h(Y_1) - \theta\})^2] \\ &= E[(\Theta - h(Y_1))^2] + E[(h(Y_1) - \theta)^2] \\ &\quad + 2E[(\Theta - h(Y_1))(h(Y_1) - \theta)].\end{aligned}$$

This last term is zero since

$$\begin{aligned}&E[(\Theta - h(Y_1))(h(Y_1) - \theta)] \\ &= \int_{-\infty}^{\infty}\int_{-\infty}^{\infty}(u - h(y_1))(h(y_1) - \theta)f_{\Theta,Y_1}(u, y_1)\,du\,dy_1 \\ &= \int_{-\infty}^{\infty}(h(y_1) - \theta)\left\{\int_{-\infty}^{\infty}(u - h(y_1))f_{\Theta|Y_1}(u \mid y_1)\,du\right\}f_{Y_1}(y_1)\,dy_1 \\ &= \int_{-\infty}^{\infty}(h(y_1) - \theta)\{0\}f_{Y_1}(y_1)\,dy_1 = 0.\end{aligned}$$

Now $\sigma_{h(Y_1)}^2 = E[h(Y_1) - \theta)^2]$ and $E[(\Theta - h(Y_1))^2] \geq 0$, so $\sigma_\Theta^2 \geq \sigma_{h(Y_1)}^2$. Notice that we have equality only when $E[(\Theta - h(Y_1))^2] = 0$; i.e., when $P(\Theta = h(Y_1)) = 1$ and the estimator Θ is a function only of the sufficient statistic, except for sets of measure 0.

The following example illustrates a use of this theorem.

Example 3.3.5

Assume X_1, X_2 is a random sample of an exponential random variable with parameter λ; thus

$$f_X(x) = \lambda e^{-\lambda x}, \qquad x > 0, \qquad \lambda > 0$$

and $E[X] = 1/\lambda$. Let $X_{(1)}$ be the minimum sample value; then

$$f_{X_{(1)}}(x) = 2\lambda e^{-2\lambda x}, \qquad x > 0$$

and $E[X_{(1)}] = 1/(2\lambda)$. Thus $2X_{(1)}$ is an unbiased estimator for the mean value of X. Now the joint density of X_1 and X_2 is

$$\lambda^2 e^{-\lambda \sum_1^2 x_i} = [\lambda^2 e^{-\lambda y}] \cdot 1$$

where $y = \sum_1^2 x_i$; thus $Y = X_1 + X_2$ is the sufficient statistic for λ. We find the joint density for $X_{(1)}$ and Y as follows: The joint density for X_1, X_2 is

$$f_{X_1 \cdot X_2}(x_1, x_2) = \lambda^2 e^{-\lambda(x_1 + x_2)}, \qquad x_1 > 0, \qquad x_2 > 0, \qquad \lambda > 0.$$

Let $x_{(1)} = \min(x_1, x_2)$, $y = x_1 + x_2$. In the half plane where $x_1 < x_2$ then $x_{(1)} = x_1$ and $y = x_1 + x_2$, $J = \begin{vmatrix} 1 & 0 \\ 1 & 1 \end{vmatrix} = 1$; in the half plane where $x_1 > x_2$, $x_{(1)} = x_2$ and $y = x_1 + x_2$, $J = \begin{vmatrix} 0 & 1 \\ 1 & 1 \end{vmatrix} = -1$. Thus in either half plane $|J| = 1$. This transformation is not 1 to 1; clearly if y and $x_{(1)}$ take on any particular values, such as 2 and 1/2, this could have occurred with either $x_1 = 1/2$, $x_2 = 3/2$, or $x_1 = 3/2$ and $x_2 = 1/2$. Thus, we must transform $f_{X_1,X_2}(x_1, x_2)$ in each of the two half planes, where $x_1 < x_2$ versus $x_1 > x_2$, and add the values of the density from these two portions together. Furthermore, since $x_{(1)} = \min(x_1, x_2)$, clearly the range for $x_{(1)}$ is 0 to $(1/2)y$ (at the boundary $x_1 = x_2$, $y = x_1 + x_2 = 2\min(x_1, x_2) = 2x_{(1)}$, otherwise $y > 2\min(x_1, x_2)$). Thus, the joint density for $X_{(1)}$ and Y is

$$f_{X_{(1)},Y}(x_1, y) = 2\lambda^2 e^{-\lambda y}, \qquad 0 < x_{(1)} < \frac{1}{2}y, \qquad y > 0.$$

The marginal density for Y is

$$f_Y(y) = \int_0^{y/2} 2\lambda^2 e^{-\lambda y}\, du = \lambda^2 y e^{-\lambda y}, \qquad y > 0$$

and the conditional density for $X_{(1)}$, given $Y = y$, is

$$f_{X_{(1)}|Y}(x_{(1)} \mid y) = \frac{2\lambda^2 e^{-\lambda y}}{\lambda^2 y e^{-\lambda y}} = \frac{2}{y}, \qquad 0 < x_{(1)} < \frac{1}{2}y.$$

Then

$$E[2X_{(1)} \mid Y = y] = \int_0^{y/2} 2x_{(1)} \frac{2}{y}\, dx_{(1)} = \frac{y}{2} = h(y),$$

and $h(Y) = \frac{1}{2}Y$ should also be an unbiased estimator for $E[X]$ and should have smaller variance than $2X_{(1)}$, by Theorem 3.3.3. It is easily verified that $E[\frac{1}{2}Y] = 1/\lambda$, $V(2X_{(1)}) = 1/\lambda^2 > 1/(2\lambda^2) = V(\frac{1}{2}Y)$.

The major use of Theorem 3.3.3, of course, is to justify the use of unbiased estimators based on sufficient statistics; we know that any unbiased estimator which is not a function of the sufficient statistic can be improved upon, in terms of its variance. Clearly, if we have the choice of two or more unbiased estimators, we would want to choose the one with the smallest variance, since we would expect its value to be closest to the true unknown parameter value, for an arbitrary sample.

A question may then arise about the unbiased estimator with the smallest possible variance. We have seen in Theorem 3.3.3 that if Θ and $h(Y_1)$ are both unbiased estimators, and Y_1 is the sufficient statistic for θ, then $\sigma^2_{h(Y_1)} \leq \sigma^2_{\Theta}$. Suppose there is a second function, $g(Y_1)$, which is also an unbiased estimator for θ. Which of $g(Y_1)$ and $h(Y_1)$ will have the smaller variance? And, indeed, are $g(Y_1)$ and $h(Y_1)$ actually different estimators? If it happens that

$$\int_{R_{Y_1}} v(y) f_{Y_1}(y)\, dy = 0$$

for each possible θ implies that $v(y) \equiv 0$, for $y \in R_{Y_1}$, then Y_1 is called a *complete sufficient statistic* for θ. In this case the unbiased estimator of θ, which is a function of the sufficient statistic, is unique and can be shown to be the *minimum variance unbiased estimator* for θ. Note that completeness is a property of $f_{Y_1}(y)$, the density for Y_1; the thorough investigation of complete density functions is beyond the scope of our discussion.

Perhaps the primary reason that maximum likelihood estimators are preferred over those generated by other methods is that, if a sufficient statistic for a parameter exists, then the maximum likelihood estimator is a function of that statistic. This result is proved below as Theorem 3.3.4.

THEOREM 3.3.4. Assume that Y is a sufficient statistic for θ based on a random sample of a random variable with density $f_X(x)$. Then the maximum likelihood estimator for θ is a function of Y.

Proof: The likelihood function for the sample is $L(\theta) = \prod_{i=1}^{n} f_X(x_i)$. If Y_1 is the sufficient statistic for θ, and $Y_2, Y_3, \ldots, Y_n$ are any additional statistics giving a nonzero jacobian for the transformation from $X_1, X_2, \ldots, X_n$ to $Y_1, Y_2, \ldots, Y_n$, then

$$\prod f_X(x_i)\,|J| = f_Y(y) f_{Y_2,\ldots,Y_n|Y_1}(y_2, \ldots, y_n \mid y),$$

where J and $f_{Y_2,\ldots,Y_n|Y_1}$ do not depend on θ; thus, if we maximize $\prod f_X(x_i)$ with respect to θ, it is equivalent to maximizing $f_{Y_1}(y)$ with respect to θ, since $f_{Y_2,\ldots,Y_n|Y_1}$ does not depend on θ. The value which maximizes $f_{Y_1}(y)$ is, of course, $h(y)$, a function of y, if we ignore trivial (constant) solutions, and thus the maximum likelihood estimator is $h(Y_1)$, a function of the sufficient statistic. The result is also easily proved for a sample of a discrete random variable.

The mean square error of an estimator, defined below, is useful in examining estimators that are not necessarily unbiased.

DEFINITION 3.3.4. If Θ is an estimator for θ, then the *mean square error* of Θ is

$$E[(\Theta - \theta)^2] = MSE(\Theta).$$

Notice immediately then,

$$E[(\Theta - \theta)^2] = E[(\Theta - E[\Theta] + E[\Theta] - \theta)^2]$$
$$= \sigma_\Theta^2 + b_\Theta^2(\theta),$$

and if Θ is unbiased, then $MSE(\Theta) = \sigma_\Theta^2$. If we can find an estimator which is biased, yet has a smaller mean square error than does an unbiased estimator, then we would possibly like to use the biased estimator. The following example illustrates a case in which the mean square error of a biased estimator is smaller than the variance of the unbiased estimator based on the sufficient statistic for the parameter.

Example 3.3.6 (*R. N. Forrest*)
Assume that X_1, X_2 is a random sample of an exponential random variable with parameter λ, as in Example 3.3.5. We saw there that the unbiased estimator of the mean of X, based on the sufficient statistic $Y = X_1 + X_2$, was $\frac{1}{2}Y = \bar{X}$ and that this estimator has variance $1/(2\lambda^2)$. As an alternative to $\bar{X}$ consider the geometric mean of X_1 and X_2, $U = \sqrt{X_1 X_2}$. Since X_1 and X_2 are independent, we have

$$E[U] = E[\sqrt{X_1}]E[\sqrt{X_2}] = \frac{\pi}{4\lambda}$$

because

$$\int_0^\infty \sqrt{x}\,\lambda e^{-\lambda x}\,dx = \frac{\sqrt{\pi}}{2\sqrt{\lambda}}.$$

Also

$$E[U^2] = E[X_1]E[X_2] = \frac{1}{\lambda^2};$$

thus

$$MSE(U) = \sigma_U^2 + b_U^2\left(\frac{1}{\lambda}\right)$$
$$= \left\{\frac{1}{\lambda^2} - \frac{\pi^2}{16\lambda^2}\right\} + \left\{\frac{\pi}{4\lambda} - \frac{1}{\lambda}\right\}^2$$
$$= \frac{4 - \pi}{2}\,\frac{1}{\lambda^2}.$$

Note that $(4 - \pi)/2 < 1/2$, and thus

$$MSE(U) < \sigma_{\bar{Y}}^2.$$

We can still do better than U, though, in estimating $E[X] = 1/\lambda$, by applying the following theorem, which you are asked to prove in Exercise 6 below.

THEOREM 3.3.5. Let $X_1, X_2, \ldots, X_n$ be a random sample of a random variable with density which depends on θ. Let Θ be a statistic and define $h(y) = E[\Theta \mid Y = y]$, where Y is a sufficient statistic for θ. Then

$$MSE(h(Y)) \leq MSE(\Theta).$$

Thus, even if we want to find estimators with small mean square error, we should make them functions of the sufficient statistic.

The Cramér-Rao inequality, which we shall now derive, gives a lower bound on the value of the mean square error or the variance of an estimator based on a sample of a continuous random variable which has a density function that is "regular" in certain ways. A similar result holds for the discrete case.

THEOREM 3.3.6. Suppose $X_1, X_2, \ldots, X_n$ is a random sample of a random variable with density $f_X(x)$, where $f_X(x)$ depends on a parameter θ and is a density for all θ in a nondegenerate open interval Ω; the range of X, $R_X = \{x\colon\ f_X(x) > 0\}$, is assumed independent of θ. Let Θ be an estimator for θ, with density $f_\Theta(u)$, and mean value $E[\Theta] = \theta + b_\Theta(\theta)$; we also assume that we can interchange integration and differentiation in certain integrals. Then

$$MSE(\Theta) \geq \frac{(1 + b'_\Theta(\theta))^2}{nE[(\log f_X(X))^2]}.$$

Proof:

$$\int_{-\infty}^{\infty} f_\Theta(u)\, du = 1, \qquad \int_{-\infty}^{\infty} u f_\Theta(u)\, du = \theta + b_\Theta(\theta)$$

and thus

$$\int_{-\infty}^{\infty} \frac{\partial}{\partial\theta} f_\Theta(u)\, du = \int_{-\infty}^{\infty} \frac{\partial \log f_\Theta(u)}{\partial\theta} f_\Theta(u)\, du = 0$$

$$\int_{-\infty}^{\infty} u\, \frac{\partial f_\Theta(u)}{\partial\theta}\, du = \int_{-\infty}^{\infty} u\, \frac{\partial \log f_\Theta(u)}{\partial\theta} f_\Theta(u)\, du = 1 + b'_\Theta(u),$$

where $\partial b_\Theta(\theta)/\partial\theta = b'_\Theta(\theta)$, assuming that $f_\Theta(u)$ and $uf_\Theta(u)$ are such that integration and differentiation can be interchanged. We then can write

$$\int_{-\infty}^{\infty} (u - \theta)\, \frac{\partial \log f_\Theta(u)}{\partial\theta} f_\Theta(u)\, du$$
$$= \int_{-\infty}^{\infty} \{(u - \theta)\sqrt{f_\Theta(u)}\}\left\{\frac{\partial \log f_\Theta(u)}{\partial\theta}\sqrt{f_\Theta(u)}\right\} du = 1 + b'_\Theta(\theta),$$

by the continuous version of the Schwartz inequality (see problem 10 below); then

$$\int_{-\infty}^{\infty} (u - \theta)^2 f_\Theta(u)\, du \int_{-\infty}^{\infty} \left(\frac{\partial \log f_\Theta(u)}{\partial \theta}\right)^2 f_\Theta(u)\, du \geq (1 + b'_\Theta(\theta))^2.$$

(When would we have equality here?) That is

$$MSE(\Theta) \geq \frac{(1 + b'_\Theta(\theta))^2}{\displaystyle\int_{-\infty}^{\infty} \left(\frac{\partial \log f_\Theta(u)}{\partial \theta}\right)^2 f_\Theta(u)\, du};$$

it remains then for us to show that

$$\int_{-\infty}^{\infty} \left(\frac{\partial \log f_\Theta(u)}{\partial \theta}\right)^2 f_\Theta(u)\, du \leq nE[(\log f_X(X))^2]$$

and the proof will be complete. We know that if $\mathbf{Y}' = (Y_2, Y_3, \ldots, Y_n)$ is any additional set of statistics such that the jacobian of the transformation from $x_1, x_2, \ldots, x_n$ to u, $\mathbf{y}'$ is nonzero, then

$$\prod_{i=1}^{n} f_X(x_i)\, |J| = f_\Theta(u) f_{\mathbf{Y}|\Theta}(\mathbf{y} \mid u),$$

where $x_1, \ldots, x_n$ are expressed in terms of $u, y_2, \ldots, y_n$. Then

$$\sum_{i=1}^{n} \log f_X(x_i) + \log |J| = \log f_\Theta(u) + \log f_{\mathbf{Y}|\Theta}(\mathbf{y} \mid u)$$

and

$$\sum_{i=1}^{n} \frac{\partial \log f_X(x_i)}{\partial \theta} = \frac{\partial \log f_\Theta(u)}{\partial \theta} + \frac{\partial \log f_{\mathbf{Y}|\Theta}(\mathbf{y} \mid u)}{\partial \theta}$$

since $|J|$ does not depend on θ. Then

$$\sum_{i=1}^{n} \left(\frac{\partial \log f_X(x_i)}{\partial \theta}\right)^2 + 2 \sum_{i<j}\sum \frac{\partial \log f_X(x_i)}{\partial \theta} \frac{\partial \log f_X(x_j)}{\partial \theta}$$
$$= \left(\frac{\partial \log f_\Theta(u)}{\partial \theta}\right)^2 + \left(\frac{\partial \log f_{\mathbf{Y}|\Theta}(\mathbf{y} \mid u)}{\partial \theta}\right)^2$$
$$+ 2\left(\frac{\partial \log f_\Theta(u)}{\partial \theta}\right)\left(\frac{\partial \log f_{\mathbf{Y}|\Theta}(\mathbf{y} \mid u)}{\partial \theta}\right).$$

Again

$$\int_{-\infty}^{\infty} f_X(x_i)\, dx_i = \int_{-\infty}^{\infty} \cdots \int_{-\infty}^{\infty} f_{\mathbf{Y}|\Theta}(\mathbf{y} \mid u)\, d\mathbf{y} = 1$$

and thus

$$\int_{-\infty}^{\infty} \frac{\partial \log f_X(x_i)}{\partial \theta} f_X(x_i)\, dx_i = \int_{-\infty}^{\infty} \cdots \int_{-\infty}^{\infty} \frac{\partial \log f_{\mathbf{Y}|\Theta}(\mathbf{y} \mid u)}{\partial \theta} f_{\mathbf{Y}|\Theta}(\mathbf{y} \mid u)\, d\mathbf{y} = 0;$$

because all cross products have expectation 0, then,

$$\begin{aligned}\sum_{i=1}^{n} \int_{-\infty}^{\infty} \left(\frac{\partial \log f_X(x_i)}{\partial \theta}\right)^2 dx_i &= n \int_{-\infty}^{\infty} \left(\frac{\partial \log f_X(x)}{\partial \theta}\right)^2 f_X(x) \\ &= \int_{-\infty}^{\infty} \left(\frac{\partial \log f_\Theta(u)}{\partial \theta}\right)^2 f_\Theta(u)\, du \\ &\quad + \int_{-\infty}^{\infty} \cdots \int_{-\infty}^{\infty} \left(\frac{\partial \log f_{\mathbf{Y}|\Theta}(\mathbf{y} \mid u)}{\partial \theta}\right)^2 f_{\mathbf{Y}|\Theta}(\mathbf{y} \mid u)\, d\mathbf{y}.\end{aligned}$$

Since

$$\int_{-\infty}^{\infty} \cdots \int_{-\infty}^{\infty} \left(\frac{\partial \log f_{\mathbf{Y}|\Theta}(\mathbf{y} \mid u)}{\partial \theta}\right)^2 f_{\mathbf{Y}|\Theta}(\mathbf{y} \mid u)\, d\mathbf{y} \geq 0,$$

we have

$$n \int_{-\infty}^{\infty} \left(\frac{\partial \log f_X(x)}{\partial \theta}\right)^2 f_X(x)\, dx \geq \int_{-\infty}^{\infty} \left(\frac{\partial \log f_\Theta(u)}{\partial \theta}\right)^2 f_\Theta(u)\, du,$$

that is

$$nE\left[\left(\frac{\partial \log f_X(X)}{\partial \theta}\right)^2\right] \geq E\left[\left(\frac{\partial \log f_\Theta(\Theta)}{\partial \theta}\right)^2\right].$$

Equality holds here only if Θ is a function of the sufficient statistic for θ, because then $f_{Y|\Theta}(\mathbf{y} \mid u)$ is independent of θ; thus the proof is complete.

Notice that if Θ is an unbiased estimator for θ, then the Cramér-Rao theorem says that

$$\sigma_\Theta^2 \geq \frac{1}{nE\left[\left(\dfrac{\partial \log f_X(X)}{\partial \theta}\right)^2\right]},$$

subject to the regularity conditions needed in the derivation. Thus, it can be used to put a bound on either the mean square error or variance of an estimator. The bound is achieved if

(a) Θ is a function of the sufficient statistic for θ.
(b) $\partial \log L/\partial \theta = k(\Theta - \theta)$, where k is a constant (possibly involving θ) and L is the likelihood function.

Example 3.3.7
Let us apply this theorem to the estimation of the mean value of a normal random variable. We have seen that $\bar{X}$ is the function of the sufficient statistic for μ ($\sum X_i$), that it is unbiased and that its variance is σ^2/n. Since, for the normal density

$$\frac{\partial \log f_X(x)}{\partial \mu} = \frac{(x - \mu)}{\sigma^2},$$

notice that

$$E\left[\left(\frac{\partial \log f_X(X)}{\partial \mu}\right)^2\right] = E\left[\left(\frac{X - \mu}{\sigma^2}\right)^2\right] = \frac{\sigma^2}{\sigma^4} = \frac{1}{\sigma^2}.$$

Thus, the Cramér-Rao bound says

$$\sigma_{\bar{X}}^2 \geq \frac{1}{n\left(\dfrac{1}{\sigma^2}\right)} = \frac{\sigma^2}{n}$$

and for this case the bound is in fact achieved. Since $f_X(x)$ does satisfy the necessary regularity conditions, $\bar{X}$ is the minimum variance unbiased estimator for the mean of a normal random variable.

It is important to realize that the density function of the population from which the sample was taken must satisfy the necessary regularity conditions or the bound derived in Theorem 3.3.6 does not apply. The following example gives an unbiased estimator whose variance is smaller than the Cramér-Rao bound, because the regularity conditions are not satisfied.

Example 3.3.8
Assume the population random variable X has density $f_X(x) = e^{-(x-\alpha)}$, $x > \alpha$ and, based on a random sample of n observations of X, we want to estimate α. It is easy to see that the maximum likelihood estimator for α is $\hat{A} = \min(X_1, \ldots, X_n) = X_{(1)}$. Notice that $f_{X_1}(x) = ne^{-n(x-\alpha)}$, $x > \alpha$ and thus

$$\mu_{X_{(1)}} = \alpha + \frac{1}{n}, \qquad \sigma_{X_{(1)}}^2 = \frac{1}{n^2}.$$

Thus $\hat{A} = X_{(1)} - (1/n)$ is an unbiased estimator for α. From Theorem 3.3.6, the smallest variance for an unbiased estimator of α should be

$$\frac{1}{nE\left[\dfrac{\partial \log f_X(X)}{\partial \alpha}\right]^2} = \frac{1}{nE[1]^2} = \frac{1}{n};$$

because $f_X(x)$ does not satisfy the regularity conditions we are able to find an unbiased estimator with a smaller variance than $1/n$. It is interesting that this estimator occurs from the use of the maximum likelihood method.

We have dwelt at some length on the estimation of a single parameter. If we are interested in simultaneously estimating two or more parameters from the same distribution function, it is easy to generalize the notion of sufficient statistics; as in the case of only one parameter, it is generally advantageous to base the estimators on the sufficient statistics. If we have k unknown parameters, then $Y_1, Y_2, \ldots, Y_k$ are jointly sufficient for the k parameters, if the conditional distribution of $Y_{k+1}, \ldots, Y_n$, given values for $Y_1, Y_2, \ldots, Y_k$, is independent of the unknown parameters. Thus, the statistics $Y_1, Y_2, \ldots, Y_k$ contain all the information in the sample about the parameters. The Fisher-Neyman factorization criterion generalizes directly for the case of two or more unknown parameters.

Example 3.3.9

Suppose $X_1, X_2, \ldots, X_n$ is a random sample of a normal random variable. Then the joint density for the sample is

$$\prod f_X(x_i) = \left(\frac{1}{2\pi\sigma^2}\right)^{n/2} e^{-\frac{1}{2}\sum_{i=1}^{n}\left(\frac{x_i-\mu}{\sigma}\right)^2}$$

$$= \frac{1}{\sigma^n}\, e^{-\frac{1}{2}\left\{\frac{n(\bar{x}-\mu)^2}{\sigma^2}+\frac{(n-1)s^2}{\sigma^2}\right\}}\cdot\left(\frac{1}{2\pi}\right)^{n/2};$$

thus $\bar{X}$ and S^2 are jointly sufficient statistics for μ and σ^2, by the factorization criterion.

There are great advantages to using functions of sufficient statistics in estimating parameters. Unfortunately, in many important applied problems sufficient statistics do not exist for parameters of interest. Still, the maximum likelihood estimates of these parameters can be expected to have many desirable properties.

Let us close this chapter by giving a result about the large sample behavior of maximum likelihood estimators. The interested reader is referred to the book by Cramér (7) for the proof of this result.

THEOREM 3.3.7. Assume that the population density is $f_X(x)$ with unknown parameter θ and that certain derivatives of f_X with respect to θ exist, are finite and integrable, for all possible values of θ. Let

$$k^2 = E\left[\left(\frac{\partial \log f_X(X)}{\partial\theta}\right)^2\right]$$

and let Θ be the maximum likelihood estimator for θ. Then, in the limit as n (the sample size) increases without bound, the distribution of $k\sqrt{n}[\Theta - \theta]$ converges to the standard normal distribution.

Notice, then, that for large n the distribution of Θ is approximately that of a normal random variable with mean θ and variance

$$\frac{1}{nk^2} = \frac{1}{nE\left[\left(\dfrac{\partial \log f_X(X)}{\partial \theta}\right)^2\right]},$$

the Cramér-Rao lower bound for unbiased estimators. Thus, at least for large n, the variance of the maximum likelihood estimator converges to the smallest possible value for an unbiased estimator of θ.

EXERCISE 3.3

1. Determine whether the maximum likelihood estimator of the mean of X is (a) unbiased, (b) consistent for each case mentioned in Exercise 3.2.13.

2. What is the sufficient statistic for
(a) p in the Bernoulli distribution?
(b) λ in the Poisson distribution?
(c) p in the geometric distribution?
(d) λ in the gamma distribution?
(e) p in the negative binomial distribution?

3. What happens if the jacobian of the transformation from $x_1, x_2, \ldots, x_n$ to $y_1, y_2, \ldots, y_n$ is zero?

4. Assume X_1, X_2 are independent $N(\mu, 1)$. Since $E[X^2] = 1 + \mu^2$ it would seem plausible that X_1^2 and X_2^2 might contain some information about μ, beyond that contained in the sufficient statistic $X_1 + X_2$. Accordingly, let $Y_1 = X_1^2 + X_2^2$, $Y_2 = X_1 + X_2$ and show that

$$f_{Y_1|Y_2}(y_1 \mid y_2) = \frac{\sqrt{2y_1 - y_2^2}}{\sqrt{2\pi}}\, e^{-(1/2)\{y_1 - (y_2{}^2/2)\}}, \quad y_1 > \frac{y_2^2}{2};$$

thus, as we know, there is no reason to use any information from the sample beyond $Y_2 = y_2$ in estimating μ.

5. Assume X_1, X_2, X_3 are independent Bernoulli random variables with parameter p. Let $Y_1 = X_1 + X_2 + X_3$, $Y_2 = g_2(X_1, X_2, X_3)$, $Y_3 = g_3(X_1, X_2, X_3)$ and derive the conditional distribution of Y_2, Y_3 given $Y_1 = 0, 1, 2, 3$.

6. Prove Theorem 3.3.5.

7. In Example 3.3.6, we saw a case in which a biased estimator for the mean of X had a smaller mean square error than the variance of the unbiased estimator of $E[X]$, based on the sufficient statistic. In this example let $Y = X_1 + X_2$, $U = \sqrt{X_1 X_2}$ and show that

$$f_{U|Y}(u \mid y) = \frac{u}{y\sqrt{y^2 - 4u^2}}, \quad u < \frac{y}{2};$$

apply Theorem 3.3.5 to find a function of Y which has smaller mean square error than U. Evaluate this mean square error and show it is smaller than that of U.

8. Show that $\bar{X}$ is the minimum variance unbiased estimator for $1/\lambda$ in the exponential distribution.

9. Assume $X_1, X_2, \ldots, X_n$ is a random sample of an exponential random variable with parameter λ. $Y = \sum X_i$ is the sufficient statistic for λ. Let a be any constant and determine the mean square error of aY as an estimator for $1/\lambda$. Show that $a = 1/(n + 1)$ minimizes this mean square error and thus, even though $\bar{X} = (1/n)Y$ is the minimum variance unbiased estimator for $1/\lambda$, $n\bar{X}/(n + 1)$ has a smaller mean square error.

10. Let $a(u)$ and $b(u)$ be two functions of u which are integrable on the real line. By looking at the discriminant of the quadratic equation in v

$$\int_{-\infty}^{\infty} [a(u) + vb(u)]^2 \, du \geq 0,$$

where v is real and not a function of u, show that

$$\left\{\int_{-\infty}^{\infty} a(u)b(u) \, du\right\}^2 \leq \left\{\int_{-\infty}^{\infty} a^2(u) \, du\right\}\left\{\int_{-\infty}^{\infty} b^2(u) \, du\right\}.$$

Apply this result in Theorem 3.3.6 with $a(u) = (u - \theta)\sqrt{f_\Theta(u)}$,

$$b(u) = \frac{\partial \log f_\Theta(u)}{\partial \theta} \sqrt{f_\Theta(u)}.$$

11. $X_1, X_2, \ldots, X_n$ is a random sample of a normal random variable with mean μ, variance σ^2. Compare the variance of S^2 and the mean square error of $[(n - 1)/n]S^2$ with the Cramér-Rao lower bounds for these quantities. What is the $MSE\{[(n - 1)/(n + 1)]S^2\}$?

12. Show that S is a consistent estimator for the standard deviation of a normal random variable.

13. If $X_1, X_2, \ldots, X_n$ is a random sample of an exponential random variable X, let M be the geometric mean of $X_1, X_2, \ldots, X_n$; i.e.,

$$M = (X_1 X_2 \cdots X_n)^{1/n}.$$

Find the mean and variance of M and show that M is not a consistent estimator for $1/\lambda$.

14. For the case mentioned in question 13, show that M is better than $\overline{X}$ in estimating $1/\lambda$ for $n = 2, 3$ (by looking at mean square errors). Thus, this is an example in which an estimator, though not consistent, does behave pretty well for very small samples.

15. The *best linear unbiased estimator* of a parameter is the linear function of the sample random variables which has the smallest variance in the class of linear unbiased estimators of the parameter. Show that $\overline{X}$ has this property in estimating the population mean, when sampling from any population with finite mean and variance. (Hint: Define $Y = \sum_{i=1}^{n} a_i X_i + b$, restrict $a_1, a_2, \ldots, a_n, b$ such that $E[Y] = \mu_X$, and then minimize σ_Y^2, subject to this restriction.)

16. For the density given in Example 3.3.8, show that $X_{(1)}$ is the sufficient statistic for α.

17. Show that $\sum_{i=1}^{n} X_i$ is the sufficient statistic for the parameter λ of the Poisson probability function, for the parameter p of the geometric probability function, and for the parameter p of the Bernoulli probability function. What is the minimum variance unbiased estimator for the parameter in each of these cases?

18. Assume we have available k independent samples, of size n_i, $i = 1, 2, \ldots, k$, from the same normal population. Let $Y = \sum_{i=1}^{k} a_i \overline{X}_i$. What values should the a_i have if Y is to be an unbiased estimator of μ with smallest possible variance?

19. Given a sample of n observations of a uniform random variable on $(0, \theta)$, show that $X_{(n)} = \max(X_1, X_2, \ldots, X_n)$ is a sufficient statistic for θ. Find the unbiased, consistent estimator $h(X_{(n)})$ of θ.

20. Given a sample of size n of a normal random variable, what is the large sample distribution of $\overline{X}$? of $[(n - 1)/n]S^2$? of S^2?

21. Given a sample of size n of a Bernoulli random variable with parameter p, what is the large sample distribution of $\overline{X}$?

22. Given a sample of a Poisson random variable with parameter λ, what is the large sample distribution of $\overline{X}$?

23. Given a random sample of a geometric random variable, what is the large sample distribution of $\overline{X}$? of $1/\overline{X}$?

CHAPTER 4

Confidence Sets

4.1 Introduction

In Chapter 3 we began our study of estimation of parameters by discussing some methods of point estimation and some properties of estimators which are at times useful in comparing estimators. If we use a point estimator of an unknown parameter we gain some idea of the numerical value of the parameter. But, if we do nothing more, we have no idea how confident we might be, given our particular sample, that our guess is anywhere close to the actual value of the parameter. If we have based our estimator on a sufficient statistic, and it is unbiased, then we may be using the unbiased estimator with the smallest possible variance, and thus, on the average, it should be the estimator to have its numerical value close to the unknown parameter value. This still does not indicate, for the given sample, how confident we may be that it is within a given distance of the unknown value.

In this chapter we shall discuss confidence intervals for parameters; these provide both an idea of the numerical magnitude of the parameter and of how confident we may be that a given interval includes the true unknown parameter value. The definition of a *confidence interval* follows.

DEFINITION 4.1.1. Assume $X_1, X_2, \ldots, X_n$ is a sample of a random variable X whose distribution depends on an unknown parameter θ. If L_1 and L_2 are statistics such that

$$P[L_1 \leq \theta \leq L_2] \geq 1 - \alpha$$

for all possible values of θ, then the interval (L_1, L_2) is a $100(1 - \alpha)\%$ *confidence interval* for θ. (Either of L_1 or L_2 may be trivial functions of the elements of the sample, that is, they may be constants, including $-\infty$ and ∞.)

Notice that the probability statement contained in this definition is radically different than those usually encountered in probability theory.

It is more usual to say $P(a \leq X \leq b) \geq 1 - \alpha$, where a and b are constants and X, the quantity in the middle, is the random variable. In $P(L_1 \leq \theta \leq L_2) \geq 1 - \alpha$, these two roles are reversed; the quantity in the middle is a constant, the unknown value of the parameter, and either or both of L_1 and L_2 are the random variables, the quantities that will vary from one sample to another. If in fact we have L_1 and L_2 (statistics) such that $P(L_1 \leq \theta \leq L_2) \geq 1 - \alpha$, then, if we were repeatedly to take samples of the same size, for each sample compute l_1 and l_2, the observed values of L_1 and L_2, we would find that at least the proportion $1 - \alpha$ of the computed intervals did include θ while no more than the proportion α did not. Given one particular sample and the observed values l_1 and l_2 we have no idea whether we have in fact bracketed θ; clearly we either do bracket it or we do not. But, since the two random variables L_1 and L_2 will bracket θ with probability at least $1 - \alpha$, we say we are at least $100(1 - \alpha)\%$ confident that this particular sample has led to an interval that brackets θ. In the next section we shall see several of the most commonly used confidence intervals.

At times it is of interest to give a region of values, in a higher dimensional space, that we can say has a given probability of simultaneously including the unknown values of several parameters. An example of this sort would be a region for μ and σ (or σ^2) for a normal random variable, which lies somewhere in the half plane $(-\infty < \mu < \infty, \sigma > 0)$ of possible values for μ and σ. The definition of a *confidence region* follows.

DEFINITION 4.1.2. Assume we have a random sample of a random vector $\mathbf{X}$ ($\mathbf{X}$ may be 1×1) and that the distribution of $\mathbf{X}$ depends upon a vector $\boldsymbol{\theta}$ of parameters. If $\mathbf{R}$ is a region in the space of possible values for $\boldsymbol{\theta}$, whose limits depend in some way on the sample values of $\mathbf{X}$, and

$$P(\boldsymbol{\theta} \in \mathbf{R}) \geq 1 - \alpha$$

for all possible $\boldsymbol{\theta}$, then $\mathbf{R}$ is a $100(1 - \alpha)\%$ *confidence region* for $\boldsymbol{\theta}$.

The interpretation of a confidence region is exactly the same as that of a confidence interval. Over repeated samples of the same size, the region $\mathbf{R}$ will be located at different places in the space for $\boldsymbol{\theta}$. In at least the proportion $(1 - \alpha)$ of these samples we can expect $\mathbf{R}$ to include the true unknown coordinate values for $\boldsymbol{\theta}$ and in no more than α of these samples we expect it will not.

Figure 4.1.1 presents a picture of a possible observed confidence region $\mathbf{r}$ for $\boldsymbol{\theta}$, where $\boldsymbol{\theta}$ is 2×1. Notice that we could in fact project the region onto the coordinate axes and thus derive interval statements about θ_1 and θ_2 individually (these would be confidence intervals for the corresponding parameters). While this sort of procedure certainly gives a legitimate

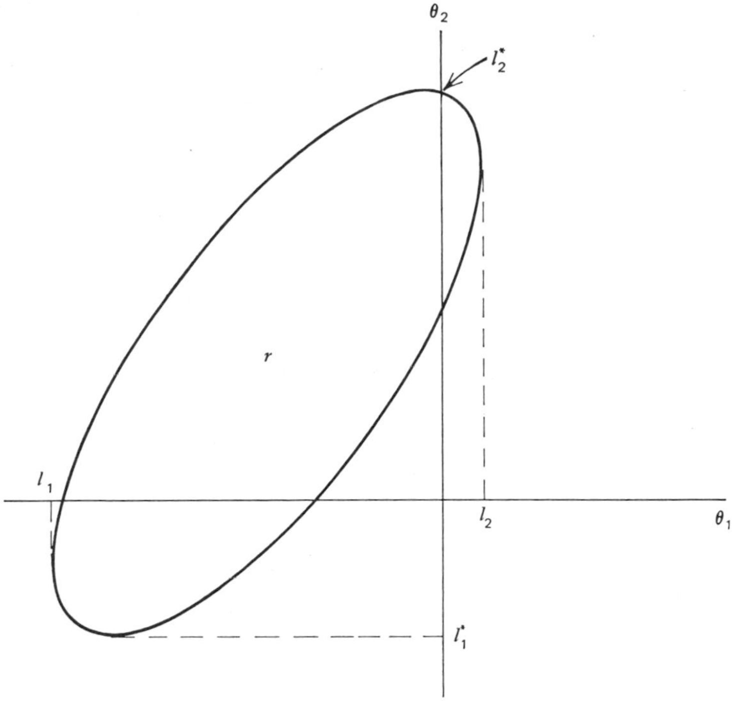

FIGURE 4.1.1. Observed confidence region r for θ.

confidence interval, it is generally not used because the resulting intervals are frequently longer than those available by using other techniques, applied for only the single parameter of interest.

EXERCISE 4.1

1. Assume that the last digits of telephone numbers in your phone book are discretely uniformly distributed over the integers 0 through 9. That is, if X is a last digit selected at random, then

$$p_X(x) = \frac{1}{10}, \qquad x = 0, 1, \ldots, 9.$$

The mean of this probability function is $\mu_X = 4.5$. Notice that

$$P(2.25 \leq X \leq 9) = \frac{7}{10}.$$

But the event A: $2.25 \leq X \leq 9$ is equivalent to $\frac{1}{2}X \leq 4.5 \leq 2X$ (since $2.25 \leq X$ implies $4.5 \leq 2X$ and $X \leq 9$ implies $X/2 \leq 4.5$). Thus

$$P\left(\frac{1}{2}X \leq \mu_X \leq 2X\right) = \frac{7}{10}$$

and $L_1 = \frac{1}{2}X$, $L_2 = 2X$ provide a 70% confidence interval for μ_X. Select a sample of 100 last digits (randomly if possible) from your phone book and, for each selected digit x, compute $l_1 = \frac{1}{2}x$, $l_2 = 2x$. In how many of your samples did (l_1, l_2) include $\mu_X = 4.5$?

2. Using the 100 digits from problem 1, pair them together sequentially, to form 50 pairs of numbers, each pair having probability function

$$p_{X_1,X_2}(x_1, x_2) = \frac{1}{100}, \qquad x_i = 0, 1, \ldots, 9, \qquad i = 1, 2.$$

Then since X_1 and X_2 are independent,

$$P\left(\frac{1}{2}X_1 \leq \mu_{X_1} \leq 2X_1, \frac{1}{2}X_2 \leq \mu_{X_2} \leq 2X_2\right) = \left(\frac{7}{10}\right)^2.$$

Thus, for any given observed pair (x_1, x_2), we have the observed confidence region with corners $(\frac{1}{2}x_1, \frac{1}{2}x_2)$, $(\frac{1}{2}x_1, 2x_2)$, $(2x_1, \frac{1}{2}x_2)$, $(2x_1, 2x_2)$. For your 50 pairs (x_1, x_2), how many include the true pair of means $(4.5, 4.5)$?

3. Given X is an exponential random variable with parameter λ, what is the probability that the interval $(\frac{1}{2}x, \infty)$ includes λ?

4. If Y is uniform on the interval $(0, b)$, compute the probability that the interval $(Y, Y/\alpha)$ brackets b (where $1/\alpha > 1$).

5. Assume Y_1 is uniform on the interval $(0, a)$, Y_2 is uniform on the interval $(0, b)$, and Y_1 and Y_2 are independent. Thus

$$f_{Y_1,Y_2}(y_1, y_2) = \frac{1}{ab}, \qquad 0 < y_1 < a, \qquad 0 < y_2 < b.$$

Compute the probability that the rectangle set, with corners

$$(Y_1, Y_2), \left(Y_1, \frac{Y_2}{1 - \sqrt{1 - \alpha}}\right), \left(\frac{Y_1}{1 - \sqrt{1 - \alpha}}, Y_2\right), \left(\frac{Y_1}{1 - \sqrt{1 - \alpha}}, \frac{Y_2}{1 - \sqrt{1 - \alpha}}\right),$$

includes the pair (a, b).

4.2 Some Common Confidence Sets

In this section we shall derive some of the most commonly used confidence sets (intervals for a single parameter, regions for two or more). In each of these, the reader will see that the confidence interval is based upon a function of a statistic and the unknown parameter which has a distribution that is free of unknown parameters. Then, a manipulation of the statistic and the parameter yields the desired confidence interval. Needless to say, it is not always possible to find such a function (of a parameter and a statistic) which is independent of the unknown parameter. Thus a more general technique is needed in such cases; this will be discussed in Section 4.3.

Let us first derive a confidence interval for the mean of a normal random variable.

THEOREM 4.2.1. Assume $X_1, \ldots, X_n$ is a random sample from a normal population with unknown mean μ and unknown variance σ^2. Then

$$L_1 = \bar{X} - t_{n-1}\left(1 - \frac{\alpha}{2}\right)\frac{S}{\sqrt{n}}, \qquad L_2 = \bar{X} + t_{n-1}\left(1 - \frac{\alpha}{2}\right)\frac{S}{\sqrt{n}}$$

form a $100(1 - \alpha)\%$ confidence interval for μ, where $\bar{X}$ is the sample mean, S the sample standard deviation and $t_{n-1}[1 - (\alpha/2)]$ is the $100[1 - (\alpha/2)]$ percentile of the t distribution with $n - 1$ degrees of freedom.

Proof: As we know from Theorem 2.3.2, $(\bar{X} - \mu)/(S/\sqrt{n})$ has the t distribution with $n - 1$ degrees of freedom, and the t distribution does not depend on μ. Recall that this density function is symmetric about the line $t = 0$ and thus

$$P\left[-t_{n-1}\left(1 - \frac{\alpha}{2}\right) \le \frac{\bar{X} - \mu}{S/\sqrt{n}} \le t_{n-1}\left(1 - \frac{\alpha}{2}\right)\right] = 1 - \alpha.$$

The event $-t_{n-1}[1 - (\alpha/2)] \le (\bar{X} - \mu)/(S/\sqrt{n}) \le t_{n-1}[1 - (\alpha/2)]$ is equivalent to $\bar{X} - t_{n-1}[1 - (\alpha/2)](S/\sqrt{n}) \le \mu \le \bar{X} + t_{n-1}[1 - (\alpha/2)](S/\sqrt{n})$; thus the probability of this latter event must also be $1 - \alpha$ and the result is established.

This result is very frequently used in applications, since generally we do not know the parameters of the normal population and may want to make inferences about μ. It is remarkable that such an interval can be derived, even though the variance σ^2 is also unknown. Notice that the length of the interval is $L_2 - L_1 = 2t_{n-1}[1 - (\alpha/2)](S/\sqrt{n})$, also a random variable. Since

$$E[S] = \frac{\sqrt{2}\,\Gamma\left(\frac{n}{2}\right)}{\sqrt{n-1}\,\Gamma\left(\frac{n-1}{2}\right)}\sigma,$$

the expected length of the interval is

$$2\frac{\sqrt{2}\,\Gamma\left(\frac{n}{2}\right)}{\sqrt{n(n-1)}\,\Gamma\left(\frac{n-1}{2}\right)}t_{n-1}\left(1-\frac{\alpha}{2}\right)\sigma.$$

Thus the expected length divided by σ is a function only of n, and in fact is a decreasing function of n. Thus, if we were willing to hazard a guess about the value of the unknown σ, we could determine a sample size n which would make this expected length over σ as small as desired.

Example 4.2.1

Theorem 4.2.1 can easily be adapted to provide confidence intervals about differences of mean values, if we have normal populations with equal variances. Suppose we bottle and sell milk which is provided by farmers cooperative association A. The butterfat content of the resulting quarts of milk is assumed to be normally distributed with mean μ_A and variance σ^2. Suppose we are approached by farmers cooperative association B that would also like to sell milk to us, perhaps at a lower price. But, we do not want to accept them as a supplier if their butterfat content differs much from our established product supplied by cooperative A. We are willing to assume that quarts of milk supplied by B again have butterfat content that is normally distributed with the same variance σ^2, but the mean level is μ_B, possibly much different than μ_A. We can put a confidence interval about $\mu_A - \mu_B$ by using Theorem 4.2.1 as follows: Assume we have a random sample of n_A butterfat contents of quarts from A and a random sample of n_B butterfat contents of quarts produced from B's supply. Let $\bar{X}_A$, $\bar{X}_B$ be the sample average contents from the two. Since the two σ^2's are assumed equal, a reasonable estimator of the common unknown value is

$$S^2 = \frac{1}{n_A + n_B - 2}\left\{\sum_{i=1}^{n_A}(X_{Ai} - \bar{X}_A)^2 + \sum_{j=1}^{n_B}(X_{Bj} - \bar{X}_B)^2\right\}.$$

Clearly, $[(n_A + n_B - 2)/\sigma^2]S^2$ is $\chi^2_{n_A+n_B-2}$, from the reproductive property of χ^2, and $\bar{X}_A - \bar{X}_B$ is normal with mean $\mu_A - \mu_B$ and variance $\sigma^2[(1/n_A) + (1/n_B)]$. Furthermore, $\bar{X}_A - \bar{X}_B$ and S^2 are independent. Thus

$$\frac{\bar{X}_A - \bar{X}_B - (\mu_A - \mu_B)}{\sigma\sqrt{\frac{1}{n_A} + \frac{1}{n_B}}}$$

is standard normal, and the ratio

$$\frac{\bar{X}_A - \bar{X}_B - (\mu_A - \mu_B)}{\sigma\sqrt{(1/n_A) + (1/n_B)}} \Big/ \frac{S}{\sigma} = \frac{\bar{X}_A - \bar{X}_B - (\mu_A - \mu_B)}{S\sqrt{(1/n_A) + (1/n_B)}}$$

has the t distribution with $n_A + n_B - 2$ degrees of freedom. Thus, just as in Theorem 4.2.1, $L_1 = \bar{X}_A - \bar{X}_B - t_{n_A+n_B-2}[1 - (\alpha/2)]S\sqrt{(1/n_A) + (1/n_B)}$, $L_2 = \bar{X}_A - \bar{X}_B + t_{n_A+n_B-2}[1 - (\alpha/2)]S\sqrt{(1/n_A) + (1/n_B)}$ form a $100(1 - \alpha)\%$ confidence interval for $\mu_A - \mu_B$. If this interval does not include zero, we would have some reason to suspect that either $\mu_A > \mu_B$ or $\mu_A < \mu_B$.

It is also easy to derive a confidence interval for the variance of a normal population, as derived in the theorem below.

THEOREM 4.2.2. Assume $X_1, X_2, \ldots, X_n$ is a random sample of a normal random variable with mean μ and variance σ^2. Then

$$L_1 = \frac{(n-1)S^2}{\chi^2_{n-1}\left(1 - \dfrac{\alpha}{2}\right)}, \qquad L_2 = \frac{(n-1)S^2}{\chi^2_{n-1}\left(\dfrac{\alpha}{2}\right)}$$

provide a $100(1 - \alpha)\%$ confidence interval for σ^2.

Proof: From Theorem 2.3.1 we know that $(n - 1)S^2/\sigma^2$ is χ^2_{n-1}. Thus, again we have a function of a statistic and an unknown parameter whose distribution does not depend on any unknown parameters. Thus

$$P\left[\chi^2_{n-1}\left(\frac{\alpha}{2}\right) \le \frac{(n-1)S^2}{\sigma^2} \le \chi^2_{n-1}\left(1 - \frac{\alpha}{2}\right)\right] = 1 - \alpha$$

and the event $\chi^2_{n-1}[\alpha/2] \le (n - 1)S^2/\sigma^2 \le \chi^2_{n-1}[1 - (\alpha/2)]$ is equivalent to

$$\frac{(n-1)S^2}{\chi^2_{n-1}\left(1 - \dfrac{\alpha}{2}\right)} \le \sigma^2 \le \frac{(n-1)S^2}{\chi^2_{n-1}\left(\dfrac{\alpha}{2}\right)};$$

thus the result is established.

Again, then, even though both μ and σ^2 are unknown, we are able to derive a confidence interval for σ^2. The length of this confidence interval is

$$(n-1)S^2\left[\frac{1}{\chi^2_{n-1}\left(\dfrac{\alpha}{2}\right)} - \frac{1}{\chi^2_{n-1}\left(1 - \dfrac{\alpha}{2}\right)}\right],$$

again a random variable, with expected value

$$(n-1)\sigma^2\left[\frac{1}{\chi^2_{n-1}\left(\frac{\alpha}{2}\right)}-\frac{1}{\chi^2_{n-1}\left(1-\frac{\alpha}{2}\right)}\right].$$

Notice this expected length does not depend on the value of μ.

Given samples from two different normal populations, as mentioned in Example 4.2.1, we can derive a confidence interval for the ratio of the two variances as follows.

THEOREM 4.2.3. Assume X_{1i}, $i = 1, 2, \ldots, n_1$, is a random sample from a normal population with mean μ_1, variance σ_1^2, and $X_{2j}, j = 1, 2, \ldots, n_2$, is a random sample from a normal population with mean μ_2, variance σ_2^2, and the two samples are independent. Then

$$L_1 = \frac{S_2^2}{S_1^2}F\left(\frac{\alpha}{2}\right), \qquad L_2 = \frac{S_2^2}{S_1^2}F\left(1-\frac{\alpha}{2}\right)$$

form a $100(1-\alpha)\%$ confidence interval for σ_2^2/σ_1^2, where both percentiles are from the F distribution with $n_1 - 1$, $n_2 - 1$ degrees of freedom and $(n_1 - 1)S_1^2 = \sum_{i=1}^{n_1}(X_{1i} - \bar{X}_1)^2$, $(n_2 - 1)S_2^2 = \sum_{j=1}^{n_2}(X_{2j} - \bar{X}_2)^2$.

The proof of this theorem is left to the reader.

The following theorem gives a confidence interval for the parameter of an exponential random variable.

THEOREM 4.2.4. Assume $X_1, X_2, \ldots, X_n$ is a random sample of an exponential random variable with parameter λ. Then $L_1 = \chi^2_{2n}[\alpha/2]/2n\bar{X}$ and $L_2 = \chi^2_{2n}[1 - (\alpha/2)]/2n\bar{X}$ form a $100(1-\alpha)\%$ confidence interval for λ.

Proof: $n\bar{X} = \sum_1^n X_i$ is a gamma random variable with parameters n and λ. Then $2\lambda n\bar{X}$ is a gamma random variable with parameters n and $1/2$; that is, it is a χ^2 random variable with $2n$ degrees of freedom. The theorem then follows directly.

Example 4.2.2

The above result has applications in getting a confidence interval for the reliability of an item whose time to failure is an exponential random variable. Thus, suppose a particular item has time to failure X, which is an exponential random variable with parameter λ. Then, its *reliability* for a fixed length of time t is defined to be $R(t) = P(X > t) = 1 - F_X(t) = e^{-\lambda t}$. Thus, note that $L_1 \le \lambda \le L_2$ is equivalent to

$$e^{-L_2 t} \le e^{-\lambda t} = R(t) \le e^{-L_1 t};$$

and $e^{-L_2 t}$, $e^{-L_1 t}$ form a $100(1-\alpha)\%$ confidence interval for $R(t)$, if L_1 and L_2 form a confidence interval for λ. For obvious reasons, generally $L_1 = 0$ and only a *one-sided interval* is desired for $R(t)$.

Let us close this section with a confidence region for the two parameters of a normal population.

THEOREM 4.2.5. Assume $X_1, X_2, \ldots, X_n$ is a random sample of a normal random variable X with mean μ and variance σ^2. Then the region in the (μ, σ) space bounded by $\mu = \bar{X} \pm z[(1+\sqrt{1-\alpha})/2]\sigma/\sqrt{n}$,

$$\sigma = \frac{\sqrt{n-1}\,S}{\sqrt{\chi^2_{n-1}\left(\dfrac{1+\sqrt{1-\alpha}}{2}\right)}}, \qquad \sigma = \frac{\sqrt{n-1}\,S}{\sqrt{\chi^2_{n-1}\left(\dfrac{1-\sqrt{1-\alpha}}{2}\right)}}$$

is a $100(1-\alpha)\%$ confidence region for (μ, σ).

Proof: As we know from Theorem 2.3.1 $\bar{X}$ and $(n-1)S^2$ are independent. Thus, the probability that $\bar{X}$ lies in an interval A, and, at the same time S^2 (or S) lies in an interval B, is given by the product of their respective probabilities. $[\sqrt{n}(\bar{X}-\mu)]/\sigma$ is $N(0, 1)$ and thus

$$P\left[-z\left(\frac{1+\sqrt{1-\alpha}}{2}\right) \leq \frac{\sqrt{n}(\bar{X}-\mu)}{\sigma} \leq z\left(\frac{1+\sqrt{1-\alpha}}{2}\right)\right] = \sqrt{1-\alpha},$$

where $z[(1+\sqrt{1-\alpha})/2]$ is the $100[(1+\sqrt{1-\alpha})/2]$ percentile of the $N(0, 1)$ distribution; that is,

$$p\left[\bar{X} - \frac{\sigma}{\sqrt{n}} z\left(\frac{1+\sqrt{1-\alpha}}{2}\right) \leq \mu \leq \bar{X} + \frac{\sigma}{\sqrt{n}} z\left(\frac{1+\sqrt{1-\alpha}}{2}\right)\right] = \sqrt{1-\alpha}.$$

Similarly $[(n-1)S^2]/\sigma^2$ is χ^2_{n-1}, so

$$P\left[\chi^2_{n-1}\left(\frac{1-\sqrt{1-\alpha}}{2}\right) \leq \frac{(n-1)S^2}{\sigma^2} \leq \chi^2_{n-1}\left(\frac{1+\sqrt{1-\alpha}}{2}\right)\right] = \sqrt{1-\alpha},$$

or

$$P\left[\frac{S\sqrt{n-1}}{\sqrt{\chi^2_{n-1}\left(\dfrac{1+\sqrt{1-\alpha}}{2}\right)}} \leq \sigma \leq \frac{S\sqrt{n-1}}{\sqrt{\chi^2_{n-1}\left(\dfrac{1-\sqrt{1-\alpha}}{2}\right)}}\right] = \sqrt{1-\alpha}.$$

Because of the independence of $\bar{X}$ and S^2

$$P\left[\bar{X} - \frac{\sigma}{\sqrt{n}} z \leq \mu \leq \bar{X} + \frac{\sigma}{\sqrt{n}} z, \frac{S\sqrt{n-1}}{\sqrt{\chi_1^2}} \leq \sigma \leq \frac{S\sqrt{n-1}}{\sqrt{\chi_2^2}}\right]$$
$$= (\sqrt{1-\alpha})^2 = 1 - \alpha,$$

where

$$z = z\left(\frac{1 + \sqrt{1-\alpha}}{2}\right), \qquad \chi_1^2 = \chi_{n-1}^2\left(\frac{1 + \sqrt{1-\alpha}}{2}\right),$$
$$\chi_2^2 = \chi_{n-1}^2\left(\frac{1 - \sqrt{1-\alpha}}{2}\right),$$

and the result is established.

The region described in this theorem is given in Figure 4.2.1. Notice that for any given sample values of $\bar{X}$ and S, we can draw in the straight lines

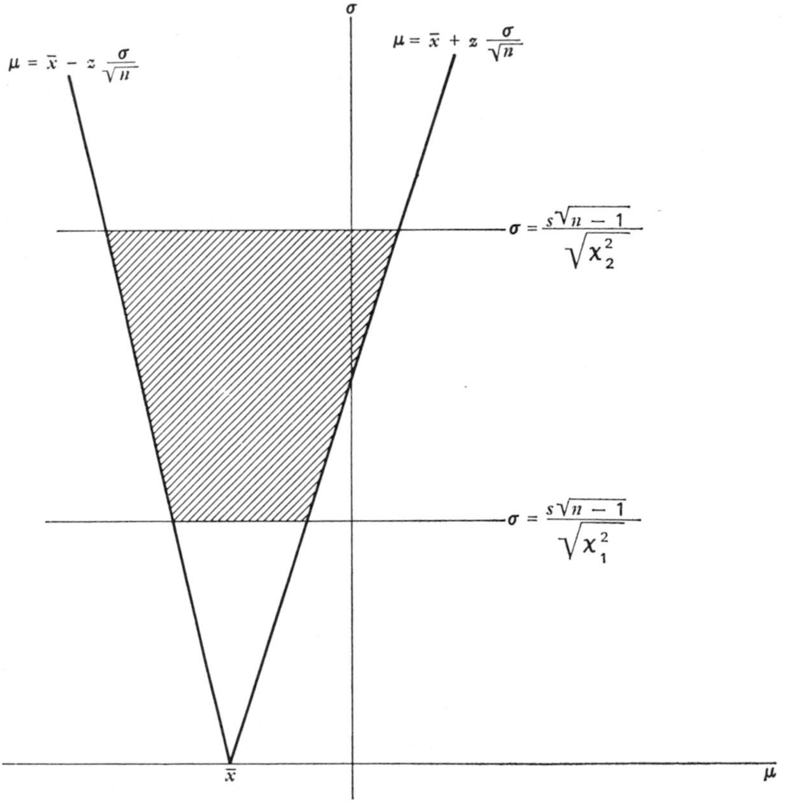

FIGURE 4.2.1. Confidence region for (μ, σ), given $\bar{x}$, s.

described above, and we have confidence $1 - \alpha$ that the resulting region includes the true unknown values of the pair (μ, σ). From one sample to the next, this trapezoid is located at a different place in the (μ, σ) plane. For $100(1 - \alpha)\%$ of the samples drawn, the region includes the true unknown values for (μ, σ). All of the confidence intervals and regions discussed in this section have been for parameters of the distributions of continuous random variables; in such cases, the probability of bracketing the unknown parameter(s) is generally exactly $1 - \alpha$. In the next section we shall discuss confidence sets for the parameters of discrete distributions, where the probability of coverage is at least $1 - \alpha$, rather than identically $1 - \alpha$.

EXERCISE 4.2

1. Given a random sample of a normal random variable with σ known, derive a $100(1 - \alpha)\%$ confidence interval for μ.

2. In Theorem 4.2.1, there are an infinite number of pairs of percentiles from the t distribution, which include the proportion $1 - \alpha$ of the density between them. In fact, if $0 < \varepsilon_1 < \alpha$, then $P(t_{n-1}(\alpha - \varepsilon_1) \leq T \leq t_{n-1}(1 - \varepsilon_1)) = 1 - \alpha$; why was $\varepsilon_1 = \alpha/2$ chosen in the statement of the theorem?

3. For a sample of a normal random variable, derive a one-sided confidence interval for μ, both for the case σ known and σ unknown; that is, find both L_1 and L_2, when σ is known and unknown, such that

$$\text{(a)}\ P(L_1 < \mu) = 1 - \alpha$$
$$\text{(b)}\ P(\mu < L_2) = 1 - \alpha.$$

4. As mentioned after Theorem 4.2.1, the expected length of the $100(1 - \alpha)\%$ confidence interval for μ, over σ, is

$$\frac{2\sqrt{2}\,\Gamma\left(\dfrac{n}{2}\right)}{\sqrt{n(n-1)}\,\Gamma\left(\dfrac{n-1}{2}\right)}\, t_{n-1}\left(1 - \frac{\alpha}{2}\right).$$

Show that this expected length converges to 0 as $n \to \infty$. If we assume $\sigma \leq 2$, how large a sample, n, do we need so that this expected length is no more than 1/4 unit, with $\alpha = .05$?

5. Again, in problem 4, assume $\sigma = 2$. What is the probability that the length is less than or equal to 1/4, for the sample size determined there?

6. For the interval given in question 1, how large a sample should be taken so that the length does not exceed 1/4 unit, with $\alpha = .05$?

7. Can we apply the technique described in Example 4.2.1 to get a confidence interval for $\mu_A - \mu_B$ if $\sigma_A \neq \sigma_B$?

8. A truck firm sends a truck each day from A to C via B. It accepts a load in A and a load in B, each day, to be delivered at C. Assume the weight of the amount loaded in A each day is well approximated by a normal random variable with mean μ_A and unknown variance σ^2 and that the amount loaded in B is well approximated by a normal random variable with mean μ_B and variance σ^2. Derive a $100(1 - \alpha)\%$ confidence interval for the mean weight on the truck when it arrives at C, given the results of n days.

9. Generalize question 8; suppose we have k normal populations with means $\mu_1, \mu_2, \ldots, \mu_k$, and that each has the same variance σ^2. Given independent random samples of size n_i, $i = 1, 2, \ldots, k$, from these populations, derive a $100(1 - \alpha)\%$ confidence interval for $\sum_{i=1}^{k} a_i\mu_i$, where $a_1, a_2, \ldots, a_k$ are any arbitrary constants.

10. Derive a $100(1 - \alpha)\%$ confidence interval for the standard deviation of a normal random variable.

11. The length of the confidence interval for σ^2, derived in Theorem 4.2.2, is proportional to $1/\chi^2[(\alpha/2)] - 1/\chi^2[1 - (\alpha/2)]$, given s^2 and n. Is this the shortest possible interval with confidence coefficient $1 - \alpha$?

12. Prove Theorem 4.2.3.

13. As in the reliability example (3.2.6), suppose we observe only the minimum value, $X_{(1)}$, of a sample of n exponential random variables. How could we use $X_{(1)}$ to construct a confidence interval for λ and for the reliability $R(t)$ of this type item for t hours (defined in Example 4.2.2)?

14. Given a random sample of size n from a normal population, how could we use Theorem 4.2.5 to get a confidence interval for t_k, the kth percentile of the normal distribution sampled from? (If X is the population random variable, $P(X \leq t_k) = k$.)

15. How could we use Theorem 4.2.5 to derive a confidence interval for the proportion of the normal population that is less than or equal to a fixed number t_0? (Thus, if $P(X \leq t_0) = p$, we want a confidence interval for p.)

16. Assume X is a uniform random variable on $(0, b)$ and that we have a random sample of n values of X. Using $X_{(n)}$, the maximum sample value, derive a $100(1 - \alpha)\%$ confidence interval for b.

17. Assume $\mathbf{X}$ is multivariate normal, $\boldsymbol{\mu}$, $\boldsymbol{\Sigma}$, and $\mathbf{X}_i$, $i = 1, 2, \ldots, n$, is a random sample of $\mathbf{X}$. Then, as we know from Exercise 2.2.17, $\overline{\mathbf{X}}$ is multivariate normal, $\boldsymbol{\mu}$, $(1/n)\boldsymbol{\Sigma}$. Then $\overline{\mathbf{X}} - \boldsymbol{\mu}$ is multivariate normal, $\mathbf{0}$, $(1/n)\boldsymbol{\Sigma}$ and $\mathbf{Y} = \sqrt{n}\,\mathbf{C}(\overline{\mathbf{X}} - \boldsymbol{\mu})$ is multivariate normal $\mathbf{0}$, $\mathbf{I}$, if $\mathbf{C}$ is any matrix such that $\mathbf{C\Sigma C}' = \mathbf{I}$. What is the distribution of $\mathbf{Y}'\mathbf{Y} = n(\overline{\mathbf{X}} - \boldsymbol{\mu})'\boldsymbol{\Sigma}^{-1}(\overline{\mathbf{X}} - \boldsymbol{\mu})$? Assuming that $\boldsymbol{\Sigma}$ is known, how could we construct a confidence region for $\boldsymbol{\mu}$, using this distribution?

18. As in problem 17, assume $\mathbf{X}_i$, $i = 1, 2, \ldots, n$, is a random sample of a multivariate normal vector $\mathbf{X}$ with $\boldsymbol{\mu}$ and $\boldsymbol{\Sigma}$. Define

$$\overline{\mathbf{X}} = \frac{1}{n}\sum_{i=1}^{n} \mathbf{X}_i, \qquad \mathbf{S} = \frac{1}{n-1}\sum_{i=1}^{n} (\mathbf{X}_i - \overline{\mathbf{X}})(\mathbf{X}_i - \overline{\mathbf{X}})'.$$

Then, it is shown in more advanced courses that

$$\frac{n(n-p)}{p}(\overline{\mathbf{X}} - \boldsymbol{\mu})'\mathbf{S}^{-1}(\overline{\mathbf{X}} - \boldsymbol{\mu})$$

is an F random variable with p, $n - p$ degrees of freedom, where p is the number of components in $\mathbf{X}$. How could we use this fact to derive a confidence region for $\boldsymbol{\mu}$, with $\boldsymbol{\Sigma}$ unknown?

19. Assume $X_1, X_2, \ldots, X_p$ is a random sample of a normal random variable with μ and σ^2. If we define $\mathbf{X}' = (X_1, \ldots, X_p)$, then $\mathbf{X}$ is multivariate normal with mean vector $\boldsymbol{\mu}' = \mu(1, 1, \ldots, 1)$ and covariance matrix $\sigma^2\mathbf{I}$. Use the result given in problem 17, $n = 1$, to get a confidence interval for μ, if σ^2 is known. Verify that this is the same interval asked for in question 1.

20. In question 18, assume that $p = 1$ and show that the confidence interval given there for μ is identical with the interval given in Theorem 4.2.1.

21. Assume that the gain X_1 in weight during the thirteenth month of a healthy, well-cared-for child is a normal random variable with mean μ_1 and variance σ_1^2; the gain X_2 in the fourteenth month is again normal, μ_2, σ_2^2 and the correlation between X_1 and X_2 is ρ, for the same child. Thus $\mathbf{X}' = (X_1, X_2)$ is bivariate normal,

$$\boldsymbol{\mu} = \begin{pmatrix} \mu_1 \\ \mu_2 \end{pmatrix}, \qquad \boldsymbol{\Sigma} = \begin{pmatrix} \sigma_1^2 & \rho\sigma_1\sigma_2 \\ \rho\sigma_1\sigma_2 & \sigma_2^2 \end{pmatrix}.$$

What is the distribution of $Y = (1, 1)\mathbf{X} = X_1 + X_2$? How could we use a random sample of $\mathbf{X}$ to get a confidence interval for $\mu_1 + \mu_2$? for $\sigma_1^2 + 2\rho\sigma_1\sigma_2 + \sigma_2^2$?

22. X is a normal random variable with $\mu = 0$ and σ^2 unknown. Given a random sample of n observations of X, derive a confidence interval for σ^2 different than the one discussed in the text. In what sense is it better than the one derived there?

4.3 A General Method

We have seen several examples of confidence intervals for parameters of continuous distributions, in which we were able to find a function of a statistic and the unknown parameter, whose distribution was independent of the unknown parameter. Then, a simple manipulation of a probability statement yielded the desired confidence interval. In many important cases it is not possible to find such a function and a different method must be sought. Let us now discuss such a method and apply it to some cases of interest.

Assume the distribution of X has one unknown parameter θ and that $\hat{\Theta}$ is an estimator of θ (a function of the sufficient statistic for θ if one exists). Then, the density or probability function for $\hat{\Theta}$ will depend on θ in some way and, in general, the $(\alpha/2)$th and $(1 - (\alpha/2))$th percentiles of this distribution will depend on θ in some way, say,

$$F_{\hat{\Theta}}(h_1(\theta)) = \frac{\alpha}{2}, \qquad F_{\hat{\Theta}}(h_2(\theta)) = 1 - \frac{\alpha}{2}.$$

Then,

$$P[h_1(\theta) < \hat{\Theta} \leq h_2(\theta)] = 1 - \alpha.$$

(Note that if X is discrete, so is $\hat{\Theta}$, in general, and it may not be possible to make $F_{\hat{\Theta}}$ equal to either $\alpha/2$ or $1 - (\alpha/2)$; in such a case we define $h_1(\theta)$ to be the largest value such that $F_{\hat{\Theta}}(h_1(\theta)) \leq \alpha/2$ and $h_2(\theta)$ to be the smallest value such that $F_{\hat{\Theta}}(h_2(\theta)) \geq 1 - (\alpha/2)$. Then $P[h_1(\theta) < \hat{\Theta} \leq h_2(\theta)] \geq 1 - \alpha/2$.) If we plot the two lines $\hat{\theta} = h_1(\theta)$, $\hat{\theta} = h_2(\theta)$ in the space for $(\hat{\theta}, \theta)$,

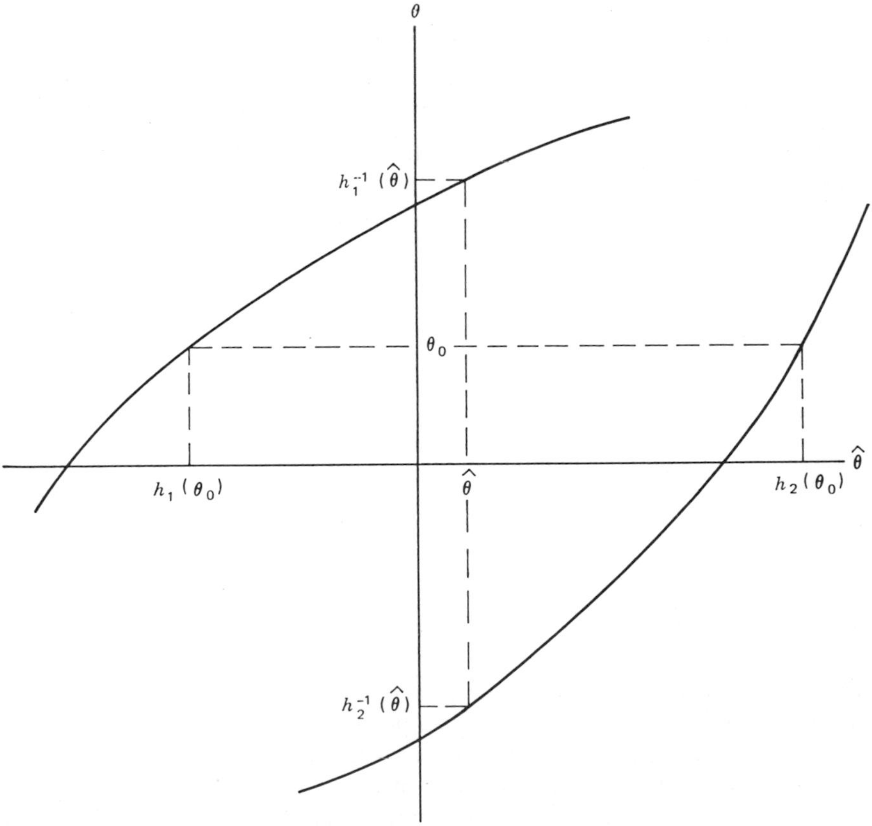

FIGURE 4.3.1. Plot of $\hat{\theta} = h_1(\theta)$, $\hat{\theta} = h_2(\theta)$.

the values for the estimate and the parameter, we have a figure similar to Figure 4.3.1. For any given value of the parameter θ, say θ_0, the probability is (at least) $1 - \alpha$ that the estimator $\hat{\Theta}$ will lie between $h_1(\theta_0)$ and $h_2(\theta_0)$ and thus in $100(1 - \alpha)\%$ of the samples we observe, $\hat{\theta}$ will lie between $h_1(\theta_0)$ and $h_2(\theta_0)$. Once we have taken the sample, and observed $\hat{\Theta} = \hat{\theta}$, we have confidence $1 - \alpha$ that the interval $[h_2^{-1}(\hat{\theta}), h_1^{-1}(\hat{\theta})]$ includes θ, since that is the probability that $\hat{\Theta}$ lies between $h_1(\theta)$ and $h_2(\theta)$. In many cases this yields a fairly easy way of constructing a confidence interval.

Example 4.3.1

Assume $X_1, X_2, \ldots, X_n$ is a random sample of a random variable X which is uniformly distributed on $(0, b)$. Then $X_{(n)}$, the largest sample value, is sufficient for b and has density

$$f_{X_{(n)}}(x) = \frac{nx^{n-1}}{b^n}, \qquad 0 \le x \le b,$$

and

$$E[X_{(n)}] = \frac{n}{n+1}\, b$$

so $\hat{B} = [(n+1)/n]X_{(n)}$ is the unbiased estimate of b which is a function of the sufficient statistic. Its density is

$$f_{\hat{B}}(u) = \frac{n^{n+1}}{b^n(n+1)^n}\, u^{n-1}, \qquad 0 \le u \le \frac{n+1}{n}\, b,$$

and

$$\int_0^{h_1(b)} f_{\hat{B}}(u)\, du = \frac{(nh_1(b))^n}{b^n(n+1)^n} = \frac{\alpha}{2},$$

$$\int_0^{h_2(b)} f_{\hat{B}}(u)\, du = \frac{(nh_2(b))^n}{b^n(n+1)^n} = 1 - \frac{\alpha}{2},$$

and we find

$$h_1(b) = \frac{n+1}{n}\, b\left(\frac{\alpha}{2}\right)^{1/n}, \qquad h_2(b) = \frac{n+1}{n}\, b\left(1 - \frac{\alpha}{2}\right)^{1/n}.$$

For any $b > 0$

$$P(h_1(b) \le \hat{B} \le h_2(b)) = 1 - \alpha;$$

inverting the functions h_1 and h_2 we find

$$h_1^{-1}(\hat{b}) = \frac{n}{n+1}\, \hat{b}\left(\frac{\alpha}{2}\right)^{-1/n} = x_{(n)}\left(\frac{\alpha}{2}\right)^{-1/n},$$

$$h_2^{-1}(\hat{b}) = \frac{n}{n+1}\, \hat{b}\left(1 - \frac{\alpha}{2}\right)^{-1/n} = x_{(n)}\left(1 - \frac{\alpha}{2}\right)^{-1/n};$$

thus

$$L_1 = X_{(n)}\left(1 - \frac{\alpha}{2}\right)^{-1/n}, \qquad L_2 = X_{(n)}\left(\frac{\alpha}{2}\right)^{-1/n}$$

form a $100(1 - \alpha)\%$ confidence interval for b.

Notice in the preceding example, as in all others, what we really want to know is $h_1^{-1}(\hat{b})$ and $h_2^{-1}(\hat{b})$, the values of b, say b_1 and b_2, such that $h_1(b_1) = h_2(b_2) = \hat{b}$; but these values are readily gotten by simply replacing $h_1(b)$ and $h_2(b)$ by $\hat{b}$ in their defining integrals (or sums) and solving the resulting equation for the unknown parameter. Thus, in the above example, we have

$$\left(\frac{n}{n+1}\right)^n \frac{1}{b^n}(\hat{b})^n = \frac{\alpha}{2}, \qquad \left(\frac{n}{n+1}\right)^n \frac{1}{b^n}\hat{b}^n = 1 - \frac{\alpha}{2}$$

which, when solved for b, yield the required functions. This technique is perhaps most useful in deriving confidence intervals for parameters of discrete distributions. In such cases the probability is at least $1 - \alpha$ that the interval covers the unknown parameter. The following example is of this type.

Example 4.3.2

Assume $X_1, X_2, \ldots, X_n$ is a random sample of a Bernoulli random variable with parameter p. Then $Y = \sum_{i=1}^n X_i$ is the sufficient statistic for p, Y is binomial with parameters n and p, and $\hat{P} = (1/n)Y$ is the unbiased estimator of p. Since Y is discrete so is $\hat{P}$ and

$$P(Y = y) = P\left(\hat{P} = \frac{y}{n}\right) \qquad \text{and} \qquad P(\hat{P} = k) = P(Y = nk).$$

Thus

$$p_{\hat{P}}(k) = \binom{n}{nk} p^{nk}(1 - p)^{n-nk}, \qquad k = 0, \frac{1}{n}, \frac{2}{n}, \ldots, 1.$$

Applying the general method described above to derive a confidence interval for p, we require $h_1(p)$ and $h_2(p)$ defined by

$$\sum_{nk=0}^{nh_1(p)} \binom{n}{nk} p^{nk}(1 - p)^{n-nk} \le \frac{\alpha}{2}$$

$$\sum_{nk=nh_2(p)}^{n} \binom{n}{nk} p^{nk}(1 - p)^{n-nk} \le \frac{\alpha}{2}$$

and we then would have $P(h_1(p) < \hat{P} \le h_2(p)) \ge 1 - \alpha$. Since, in this discrete case, nh_1 and nh_2 describe the number of terms to be added together in a binomial sum, clearly they must take on integer values, and the graph described in the general method consists simply of $n + 1$ pairs of points, a value for h_1 and h_2 at each of the possible values of $\hat{p}$: $0, 1/n, \ldots, 1$. Furthermore, for any given value $\hat{p}$ we need to know only $h_2^{-1}(\hat{p})$ and $h_1^{-1}(\hat{p})$ which are the observed confidence limits for p, given the value of $\hat{p}$. As discussed above, having observed $Y = y$, $\hat{p} = y/n$, then, $h_1^{-1}(\hat{p})$ and $h_2^{-1}(\hat{p})$ are the solutions for p in

$$\sum_{nk=0}^{y} \binom{n}{nk} p^{nk}(1-p)^{n-nk} = \frac{\alpha}{2}: \qquad \text{determines } h_1^{-1}(\hat{p})$$

$$\sum_{nk=y}^{n} \binom{n}{nk} p^{nk}(1-p)^{n-nk} = \frac{\alpha}{2}: \qquad \text{determines } h_2^{-1}(\hat{p}).$$

We use equality here because in the first equation we are adding together "left-hand" probabilities from a binomial distribution and, for a fixed number of terms, y, this sum is clearly a decreasing function of p. If p is determined with equality, then any larger value of p will have this sum strictly smaller than $\alpha/2$, and we thus have the smallest value of p such that $P(Y \leq y) \geq \alpha/2$. Similar reasoning applies for the other equation. It should be remarked that we take $h_2^{-1}(0) = 0$ and $h_1^{-1}(1) = 1$; that is, if we observe $y = 0$ successes, then the lower confidence limit is taken as 0 and, if we observe all successes, the upper confidence limit is 1.

To make this reasoning clear, let us examine the case in which $n = 2$ in some detail. The range of Y is $\{0, 1, 2\}$, and the range of $\hat{P}$ is $\{0, 1/2, 1\}$. Then $h_2^{-1}(0) = 0$, and $h_1^{-1}(0)$ is the root of

$$\sum_{nk=0}^{0} \binom{2}{nk} p^{nk}(1-p)^{2-nk} = (1-p)^2 = \frac{\alpha}{2};$$

thus $h_1^{-1}(0) = 1 - \sqrt{\alpha/2}$. Similarly, we determine $h_2^{-1}(1/2)$ from

$$2p(1-p) + p^2 = -p^2 + 2p = \frac{\alpha}{2},$$

giving $h_2^{-1}(1/2) = 1 - \sqrt{1 - (\alpha/2)}$ and we determine $h_1^{-1}(1/2)$ from

$$(1-p)^2 + 2p(1-p) = 1 - p^2 = \frac{\alpha}{2}$$

to be $h_1^{-1}(1/2) = \sqrt{1 - (\alpha/2)}$. We solve

$$p^2 = \frac{\alpha}{2}$$

to find $h_2^{-1}(1) = \sqrt{\alpha/2}$, and $h_1^{-1}(1) = 1$. Then the graph of h_1, h_2 looks something like Figure 4.3.2, as long as $\alpha \leq 1/2$, which in practice it always will be. Another graph which is of interest is the plot of the probability

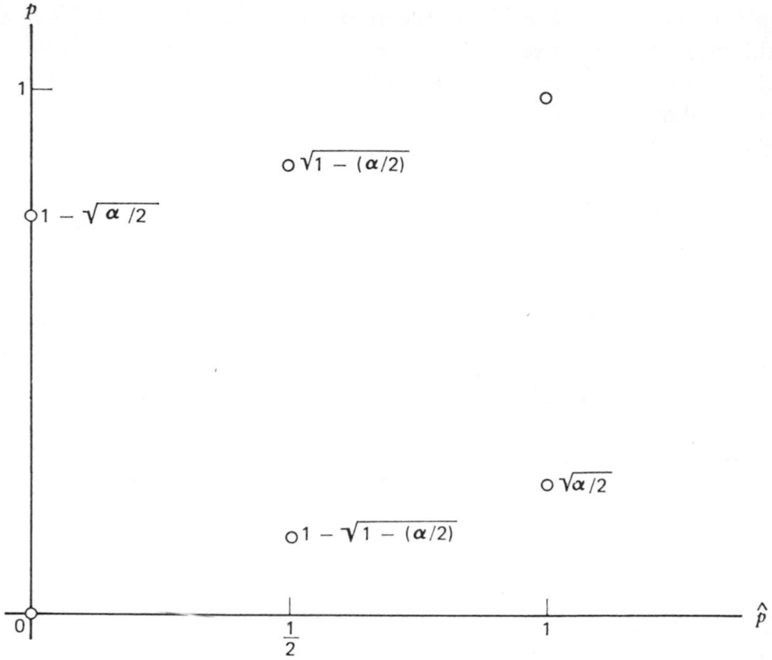

FIGURE 4.3.2. $h_1(p)$ **and** $h_2(p)$ **for binomial,** $n = 2$**.**

that the confidence interval, derived above, actually includes the true unknown value of p, as a function of p; we know from the way the interval was derived that this probability is at least $1 - \alpha$ for all p. It is quite enlightening to examine the values of p for which equality holds and to see how much greater than $1 - \alpha$ the probability of coverage can become.

It is plain that, for any given true value of p (again, for simplicity, $n = 2$),

$$P(h_2^{-1}(\hat{P}) \le p < h_1^{-1}(\hat{P})) = \sum_{k=0}^{2} P(h_2^{-1}(\hat{P}) \le p < h_1^{-1}(\hat{P}) \mid \hat{P} = k) p_{\hat{P}}(k).$$

$P(h_2^{-1}(\hat{P}) \le p < h_1^{-1}(\hat{P}) \mid \hat{P} = k)$ is either 0 or 1, of course, since either the interval does include p or it does not. Thus, the probability of coverage is simply given by adding together the probabilities that $\hat{P}$ takes on the appropriate values so that the interval does cover p. Table 4.3.1 gives the possible values of p and the corresponding values for $h_1(p)$ and $h_2(p)$. Then, to complete the computation, we need merely evaluate the corresponding probabilities that $\hat{P}$ takes on values such that the determined interval includes p. For example, if $0 \le p < 1 - \sqrt{1 - (\alpha/2)}$, the confidence interval will include p if and only if $\hat{P} = 0$ ($Y = 0$); the probability of this occurring is

TABLE 4.3.1. Values of p, $h_1(p)$, $h_2(p)$

p	$h_1(p)$	$h_2(p)$
$0 \le p < 1 - \sqrt{1 - \frac{\alpha}{2}}$	0	0
$1 - \sqrt{1 - \frac{\alpha}{2}} \le p < \sqrt{\frac{\alpha}{2}}$	0	$\frac{1}{2}$
$\sqrt{\frac{\alpha}{2}} \le p \le 1 - \sqrt{\frac{\alpha}{2}}$	0	1
$1 - \sqrt{\frac{\alpha}{2}} < p \le \sqrt{1 - \frac{\alpha}{2}}$	$\frac{1}{2}$	1
$\sqrt{1 - \frac{\alpha}{2}} < p \le 1$	1	1

$(1-p)^2$. If $1 - \sqrt{1 - (\alpha/2)} \le p < \sqrt{\alpha/2}$, then the interval will include p if and only if $\hat{P} = 0$ or $\hat{P} = 1/2$ ($Y = 0$ or $Y = 1$); the probability of this occurring is $(1-p)^2 + 2p(1-p) = 1 - p^2$. Similarly, for the remainder of the range for p, the probability that the resulting interval does include p is easily determined. If we plot this probability that the interval includes p, a picture like Figure 4.3.3 will result. Note that, as long as $\alpha \le 1/2$, then indeed $P(h_1(\hat{P}) \le p \le h_2(\hat{P})) \ge 1 - \alpha$, and, in fact, this probability exceeds $1 - (\alpha/2)$ with $n = 2$. If we take $\alpha > 1/2$ then the graph is as given in Figure 4.3.4; note that even in this unrealistic case ($\alpha > 1/2$) the probability of coverage approaches $2\sqrt{\alpha/2} - \alpha > 1 - \alpha$ (even in this case it is not attained; it is the limiting value as p approaches $1 - \sqrt{\alpha/2}$ or $\sqrt{\alpha/2}$, from $1/2$). Since, with $\alpha \le 1/2$, the probability of coverage is at least $1 - \alpha/2$, it is plain that for $n = 2$ we should replace $\alpha/2$ by α; then if $\hat{p} = 0$, the confidence interval is $(0, 1 - \sqrt{\alpha})$, for $\hat{p} = 1/2$ it is $(1 - \sqrt{1-\alpha}, \sqrt{1-\alpha})$ and for $\hat{p} = 1$ it is $(\sqrt{\alpha}, 1)$. In general, we should not replace $\alpha/2$ by α to get a $100(1-\alpha)\%$ confidence interval for p. It is just because of the special form of the binomial probability function, with $n = 2$, that this substitution still yields a $100(1-\alpha)\%$ interval for p.

An approximate procedure for getting a confidence interval for an unknown parameter, for large samples, can frequently be obtained by using the central limit theorem approximation to the distribution of the estimator. For example, in Theorem 3.3.7 we saw that if the population density is $f_X(x)$, which depends on an unknown parameter θ and satisfies certain regularity conditions, then, for large samples the maximum likelihood estimator Θ for θ has approximately a normal distribution with mean θ and variance

$$\frac{1}{nE\left[\left(\dfrac{\partial \log f_X(X)}{\partial\theta}\right)^2\right]} = \sigma_\Theta^2.$$

Thus, for large n,

$$P\left(-z_{1-\alpha/2} \leq \frac{\Theta - \theta}{\sigma_\Theta} \leq z_{1-\alpha/2}\right) \doteq 1 - \alpha,$$

where $z_{1-\alpha/2}$ is the $100[1 - \alpha/2]$ percentile of the standard normal distribution. Frequently it is fairly easy to manipulate this inequality to get an approximate confidence interval for θ. In the following example, we shall apply this

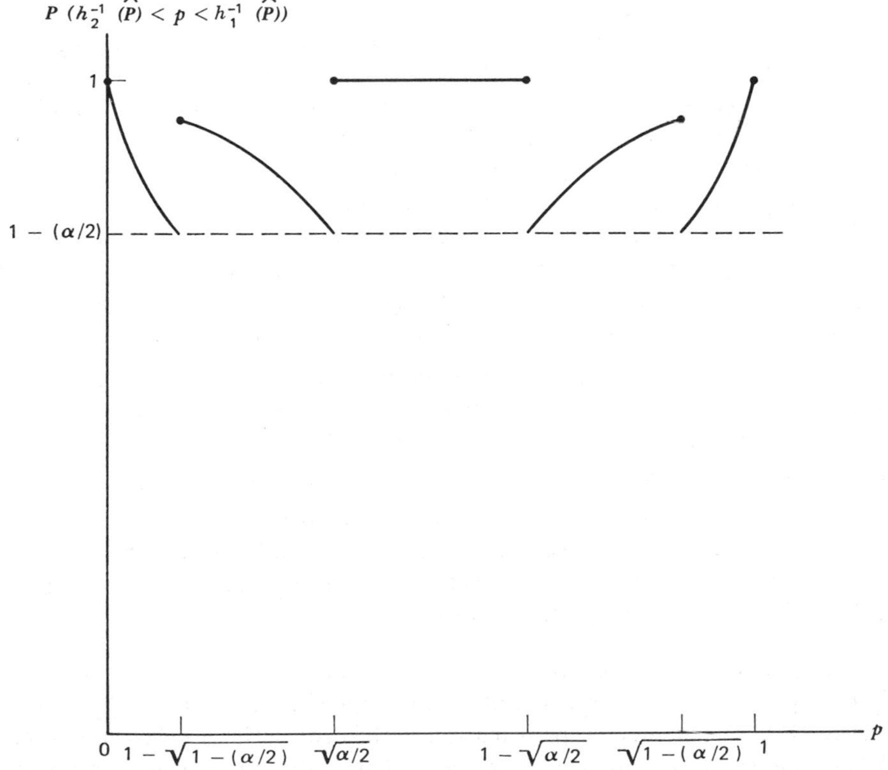

FIGURE 4.3.3. $P(h_2^{-1}(\hat{P}) < p < h_1^{-1}(\hat{P}))$ **versus** p, $n = 2$, $\alpha \leq 1/2$.

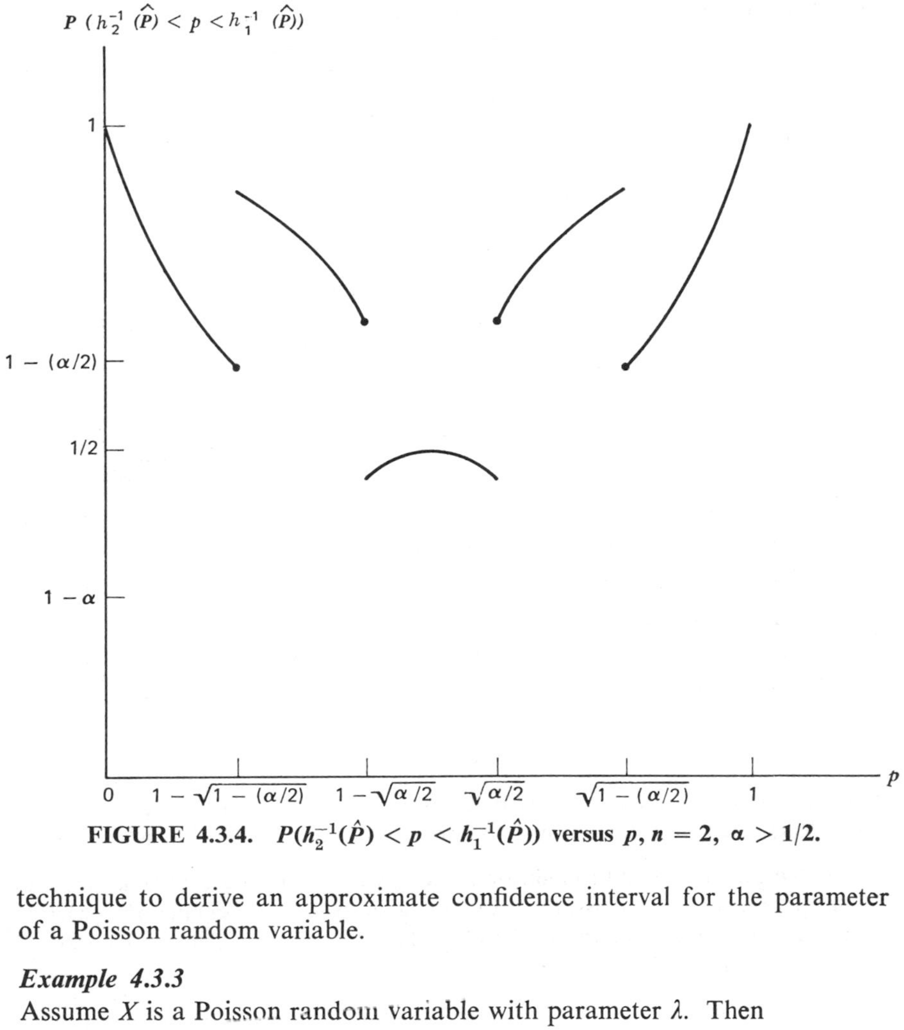

FIGURE 4.3.4. $P(h_2^{-1}(\hat{P}) < p < h_1^{-1}(\hat{P}))$ **versus** $p, n = 2, \alpha > 1/2$.

technique to derive an approximate confidence interval for the parameter of a Poisson random variable.

Example 4.3.3
Assume X is a Poisson random variable with parameter λ. Then

$$p_X(x) = \frac{\lambda^x}{x!} e^{-\lambda}$$

$$\log p_X(x) = x \log \lambda - \log x! - \lambda$$

$$\frac{\partial \log p_X(x)}{\partial \lambda} = \frac{x}{\lambda} - 1.$$

Thus

$$E\left[\left(\frac{X}{\lambda} - 1\right)^2\right] = \frac{E[X^2]}{\lambda^2} - \frac{2\lambda E[X]}{\lambda^2} + 1 = \frac{1}{\lambda}$$

and, for large samples, the maximum likelihood estimator of λ, $\bar{X}$, is approximately normal with mean λ and variance λ/n. Thus

$$P\left(-z\left(1-\frac{\alpha}{2}\right) \leq \frac{\sqrt{n}(\bar{X}-\lambda)}{\sqrt{\lambda}} \leq z\left(1-\frac{\alpha}{2}\right)\right) \doteq 1-\alpha.$$

This inequality is equivalent to

$$\frac{n(\bar{X}-\lambda)^2}{\lambda} \leq z^2\left(1-\frac{\alpha}{2}\right)$$

or

$$\lambda^2 + \lambda\left(-2\bar{X} - \frac{z^2(1-\alpha/2)}{n}\right) + \bar{X}^2 \leq 0.$$

Thus, the values of λ which make this parabola negative form the confidence interval for λ. We find, then, that $L_1 = \bar{X} + z^2/2n - \sqrt{\bar{X}z^2/n + z^4/4n^2}$ and $L_2 = \bar{X} + z^2/2n + \sqrt{\bar{X}z^2/n + z^4/4n^2}$, where $z = z(1-\alpha/2)$, form an approximate $100(1-\alpha)\%$ confidence interval for λ. Since n is assumed large, we might delete the terms in $1/n$ yielding $L_1 = \bar{X} - (z/\sqrt{n})\sqrt{\bar{X}}$, $L_2 = \bar{X} + (z/\sqrt{n})\sqrt{\bar{X}}$ as the approximate confidence interval for λ. Note that these limits are the ones we would get if $\bar{X}$ were normal with mean λ and variance $\bar{X}/n$, rather than λ/n.

EXERCISE 4.3

1. Given a random sample of n observations of a random variable which is uniform on $(-b, b)$, use the general method to derive a confidence interval for b.

2. Use the general method to indicate the way you could get a confidence interval for the parameter λ of a Poisson random variable, based on a sample of n observations. Given $n = 2$, plot the probability that the confidence interval covers λ, similar to Example 4.2.2.

3. Given a sample of size $n = 2$ of a geometric random variable, use the general method to get a confidence interval for p and plot the probability that the interval covers p.

4. Use the general method to derive a confidence interval for μ for a normal random variable with $\sigma^2 = 1$ and compare this with Exercise 4.2.1. What is the probability this interval covers μ, as a function of μ?

5. Use the general method to derive a confidence interval for σ^2 of a normal random variable.

6. Use the (approximate) asymptotic method to derive a confidence interval for p for a binomial random variable with large n.

7. Use the asymptotic method to derive an approximate confidence interval for the parameter p of a geometric random variable.

8. Given a large sample of a negative binomial random variable X with parameters r (known) and p, derive an approximate confidence interval for p; also derive an approximate confidence interval for μ_X.

9. Given a large sample of a gamma random variable X with parameters r (known) and λ unknown, derive an approximate confidence interval for λ.

10. Given a large sample of a normal random variable, derive an approximate confidence interval for σ^2. Compare this interval with the one derived in Section 4.2 for this case.

11. By referring back to Exercise 1.3.14, show how a table of the incomplete beta function could be used to compute an exact confidence interval for the parameter p of a Bernoulli random variable, given a random sample of n observations.

12. Show how a table of the incomplete gamma function can be used to compute an exact confidence interval for the parameter λ of a Poisson random variable, given a random sample of n observations (refer to Exercise 1.3.13).

[illegible]

CHAPTER 5

Tests of Hypotheses

5.1 Introduction

We have examined the concepts involved in estimating parameters, using both point and interval estimators. In this chapter, we shall examine the classical methods for testing hypotheses. First of all, a *hypothesis* is a statement about the probability distribution of a random variable (thus it can also be taken as a statement about the distribution of values in some population). A *test* of a hypothesis is simply a rule for deciding to either accept or reject the stated hypothesis. Generally, we shall want to study various rules that are based upon the possible outcomes that we might observe in taking a sample of the random variable (or from the population), since this would seem the most natural way of gathering evidence regarding the truth of the hypothesis.

In the classical framework, we shall allow ourselves only two possible decisions about the hypothesis, given the observed sample of the random variable; we shall either *accept* the hypothesis as stated or *reject* it, with a fixed sample size n of values selected from the population. (In Section 5.5 we shall study a more modern framework in which we allow ourselves three possible decisions: accept, reject, or no decision yet, take another sample.) Thus, in this classical framework, we shall partition the sample space into two parts: the *acceptance region* A and the *rejection region* (or *critical region*) $\bar{A}$. Then, if we observe a sample lying in A, we accept the hypothesis; if not, we reject it. The test or rule, then, simply specifies the way in which this partition is to be made. Two tests are different if they partition the sample space differently (thus if A_1 and A_2 are the two acceptance regions, the tests are different if and only if $A_1 \neq A_2$).

Notice that in looking at things this way there are two types of errors that we might commit. We might reject the hypothesis when it is true, called a *type I error*, or we might accept the hypothesis when it is not true, called a *type II error*. Comparisons of different tests are accomplished by comparing the sizes of these types of error. If we could find a test (partition) which simultaneously minimized the probabilities of both these types of error, it would clearly be the one we would like to use. It is easy to see that such an ideal test could not possibly exist, since for any hypothesis we could use the rule: ignore the sample and reject the hypothesis. Thus, $A = \phi$ and $\bar{A} = S$. We would then clearly have P(type II error) $= 0$, since, if we never accept the hypothesis, we cannot commit a type II error. We might also have P(type I error) $= 1$, with this procedure. If we take the complement of this procedure (that is, always accept the hypothesis), then we would have P(type I error) $= 0$ and P(type II error) $= 1$. Thus we can always find a test which makes the probability of one or the other of the two types of error 0, but we would not like to use either of these procedures, since they ignore any evidence given by the sample regarding the hypothesis.

The classical method of testing hypotheses solves this dilemma by suggesting that we consider all tests with P(type I error) $= \alpha$, and, among these, choose that one which minimizes the P(type II error). It should be mentioned that the probability of a type II error remains poorly defined at this point. To compute the probability that we accept the hypothesis when it is false, we must make some assumption about the distribution sampled, which is different than the stated hypothesis. Thus, to give concrete meaning to these two types of errors, we actually have two hypotheses in mind, the *null hypothesis* and the *alternative hypothesis*. We shall denote the null hypothesis by H_0 and the alternative hypothesis by H_1. Then, P(type I error) is computed assuming that the random variable has the distribution specified by H_0 and P(type II error) is computed under the assumption that H_1 is true. If H_0 and H_1 completely specify the population distribution then these two probabilities of error have unique values; in such a case we say that H_0 and H_1 are *simple hypotheses*. If H_1, for example, does not specify a unique distribution for the population, but rather some class of distributions, then the probability of type II error is a function, taking on a (possibly different) value for the different distributions in the class. The same is true of the probability of type I error, if H_0 specifies a class of distributions. If a hypothesis is not simple, it is called a *composite hypothesis*. The *power of a test* is the probability that we will reject H_0 (and thus accept H_1). If either H_0 or H_1 is composite then the power of the test is a function and its value can be plotted for the various possibilities specified by H_0 and H_1. Note that, assuming H_0 is true, the power of the test is the probability of a type I error, whereas, assuming H_1 is true, the power of the test is 1 minus the probability of a type II error. More exactly, if H_0 is composite, then the probability of a type I error is actually a function and α denotes the supremum of this probability, over all possible values of the parameter specified by H_0. Many people call α the *significance level* of the test. Thus, if we consider all tests with probability of type I error equal to α, we are considering all tests whose power function does not exceed α, assuming H_0 is true. We would of course like to choose that test which has the highest possible power, assuming H_1 is true.

There are two distinctly different types of hypotheses that we shall consider; the first of these is called a *parametric* hypothesis, one in which the form of the population distribution is assumed known, but the parameters are not, and the hypotheses then state the value(s) for the parameter(s) or a range of values for them. For example, if we consider a conceptual population of flies, each sprayed with a certain concentration of DDT, we can reasonably assume that each fly is either killed (1) or not (0); then if we randomly select a sample of n flies from this conceptual population, we may assume that the elements of the sample are independent Bernoulli random variables and we know the form of the distribution for the sample. But we do not know the value of the parameter p and might hypothesize that $p = 1/3$ versus $p = 1/2$; then we have stated a simple null hypothesis H_0: $p = 1/3$ versus a simple alternative hypothesis H_1: $p = 1/2$. Or we might choose H_0: $p \leq 1/3$, H_1: $p > 1/3$ as our null and alternative hypotheses, in which case both are composite. These are examples of parametric hypotheses, since they state values for unknown parameters. We shall first concern ourselves with parametric hypotheses.

The second type of hypothesis, which we shall subsequently study, are ones in which the hypothesis makes a statement about the form of the population distribution, rather than taking that as given by assumption. For example, we might hypothesize that the weights of 25-year-old males in the United States are normally distributed. Then, on the basis of a sample, we would want to accept or reject this statement about the form of the distribution sampled. The alternative hypothesis in a situation like this is much more difficult to classify, and in general it is not as straightforward to make statements about the optimality of the test or rule employed. In the case of parametric hypotheses, the alternative hypothesis still assumes that the form of the distribution function is known, but the parameter values are different from the values specified by H_0. In hypotheses about the form of the distribution, the alternatives are not so nicely bounded.

In considering parametric hypotheses we should like to accept the null hypothesis if the sample is more "consistent" with it than with H_1, in some sense. As might be expected, "consistency" with H_0 versus H_1 is essentially measured by examining the sample estimate(s) of the parameter(s) involved in the hypotheses. If the estimate(s) are "closer" to the value(s) specified by H_0 than by H_1, we would generally like to accept H_0 (reject H_1), and if the opposite is true, we would accept H_1 (reject H_0). We shall consider the methodology for testing a simple H_0 versus a simple H_1 in Section 5.2.

EXERCISE 5.1

1. It is assumed that weights of 25-year-old males are normally distributed with $\sigma = 20$ pounds. We will select a random sample of 100 people from this population and want to test H_0: $\mu = 165$ versus H_1: $\mu > 165$. Show that each of the following acceptance regions has P (type I error) = .05.

(a) Accept H_0 if we select a black card from a well-shuffled deck containing 19 black cards and 1 red card.
(b) $A = \{\bar{X}: \ \bar{X} \leq 168.28\}$
(c) $A = \{\bar{X}: \ \bar{X} \geq 161.72\}$
(d) $A = \{\bar{X}: \ 161.08 \leq \bar{X} \leq 168.92\}$

2. In problem 1, plot P (type II error) versus μ, for all $\mu > 165$. Which of the four tests would you prefer to use?

3. The time to failure of an electronic device is an exponential random variable. n of these devices are placed on test until they fail. Let $\bar{X}$ be their average time to failure and let $X_{(1)}$ be the minimum time to failure. It is desired to test H_0: $\lambda = \lambda_0$ versus H_1: $\lambda < \lambda_0$. Show that each of the following acceptance regions has P (type I error) = α. By examining the power of these tests, which would you recommend as being the best?

(a) $A = \left\{\bar{X}: \ \bar{X} \geq \dfrac{\chi^2_{2n}(\alpha)}{2n\lambda_0}\right\}$

(b) $A = \left\{\bar{X}: \ \bar{X} \leq \dfrac{\chi^2_{2n}(1-\alpha)}{2n\lambda_0}\right\}$

(c) $A = \left\{X_{(1)}: \ X_{(1)} \geq -\dfrac{1}{n\lambda_0}\log(1-\alpha)\right\}$

(d) $A = \left\{X_{(1)}: \ X_{(1)} \leq -\dfrac{1}{n\lambda_0}\log(\alpha)\right\}$

4. Assume X is a uniform random variable on the interval $(-b, b)$. We want to test H_0: $b = 1$ versus H_1: $b > 1$. We will take one observation of X and will reject H_0 if $|X| > .99$. Find the probability of type I error in using this test and plot the power function versus the values of b specified by H_1.

5. A new coin is to be flipped two times. We want to test H_0: $p = 1/2$ versus H_1: $p = 1/4$ or $p = 3/4$, where p is the probability of getting a head with this coin. If we get one head we will accept H_0, otherwise we will reject it. Find the probability of a type I error for this test and the power of the test, given $p = 1/4$ or $p = 3/4$.

6. In question 5, assume the hypotheses are H_0: $p = 1/2$ versus H_1: $p \neq 1/2$ and that the same test is used. Plot the power function in this case.

5.2 Simple Hypotheses

We shall in this section examine the special case of testing a simple hypothesis versus a simple alternative. The Neyman-Pearson fundamental lemma, Theorem 5.2.1 below, provides a methodology for finding the *best test* among all those with the same probability of type I error; best test here means that one with the lowest possible value of the probability of type II error (or alternatively, the one with the highest power).

THEOREM 5.2.1 (*Neyman-Pearson*). Assume X is a random variable whose probability law depends on an unknown parameter θ (X or θ or both may in fact be vectors). To test H_0: $\theta = \theta_0$ versus H_1: $\theta = \theta_1$, where θ_0 and θ_1 are any specified values for θ, let A (the acceptance region) for a random sample of n values of X be defined by

$$(x_1, x_2, \ldots, x_n) \in A \Leftrightarrow \frac{L(\theta_1)}{L(\theta_0)} < k$$

where $L(\theta)$ is the likelihood function of the sample and k is a constant. If $P((X_1, X_2, \ldots, X_n) \in A \mid \theta = \theta_0) = 1 - \alpha$, then this test has the smallest possible value for $\beta = P((X_1, X_2, \ldots, X_n) \in A \mid \theta = \theta_1)$, among all tests of size α (that is, among all possible partitions having the same probability of type I error).

Proof: Note first of all that the choice of the constant k determines the value of α; generally, if X is continuous we have complete freedom of choice for α because α will vary continuously with k. If, however, X is discrete, then α will vary discretely and only certain values will be attainable, just as we found for confidence coefficients in Chapter 4 in the discrete case. We shall give the proof only in the continuous case and shall let $\int_B L(\theta_0)$ and $\int_B L(\theta_1)$ denote the probability that $(X_1, X_2, \ldots, X_n) \in B \subset S$ (the sample space), assuming θ_0 and θ_1 give the true value of θ, respectively. Thus

$$\int_A L(\theta_0) = 1 - \alpha,$$

and we want to consider all other possible acceptance regions $A^* \subset S$ such that $\int_{A^*} L(\theta_0) = 1 - \alpha$; we want to establish that $\int_A L(\theta_1) \leq \int_{A^*} L(\theta_1)$ for all such A^*. Clearly we can decompose A and A^* into disjoint sets by the identities

$$A = (A \cap A^*) \cup (A \cap \bar{A}^*)$$
$$A^* = (A^* \cap A) \cup (A^* \cap \bar{A})$$

where $\bar{A}^*$ is the complement of A^* $(S - A^*)$. Because of this disjointness

$$\int_A L(\theta_1) = \int_{A \cap A^*} L(\theta_1) + \int_{A \cap \bar{A}^*} L(\theta_1)$$

and

$$\int_{A^*} L(\theta_1) = \int_{A^* \cap A} L(\theta_1) + \int_{A^* \cap \bar{A}} L(\theta_1).$$

Thus

$$\int_{A^*} L(\theta_1) - \int_A L(\theta_1) = \int_{A^* \cap \bar{A}} L(\theta_1) - \int_{A \cap \bar{A}^*} L(\theta_1).$$

Now A is the set of points $(x_1, x_2, \ldots, x_n)$ such that $[L(\theta_1)/L(\theta_0)] < k$ and thus $L(\theta_1) < kL(\theta_0)$ for $(x_1, x_2, \ldots, x_n) \in A$, $L(\theta_1) \geq kL(\theta_0)$ for $(x_1, x_2, \ldots, x_n) \in \bar{A}$. Thus

$$\int_{A^* \cap \bar{A}} L(\theta_1) \geq k \int_{A^* \cap \bar{A}} L(\theta_0)$$

and

$$\int_{A \cap \bar{A}^*} L(\theta_1) < k \int_{A \cap \bar{A}^*} L(\theta_0).$$

Subtracting the second inequality from the first gives

$$\int_{A^* \cap \bar{A}} L(\theta_1) - \int_{A \cap \bar{A}^*} L(\theta_1) \geq k \left\{ \int_{A^* \cap \bar{A}} L(\theta_0) - \int_{A \cap \bar{A}^*} L(\theta_0) \right\}.$$

Note, then, that

$$\int_{A^* \cap \bar{A}} L(\theta_0) - \int_{A \cap \bar{A}^*} L(\theta_0) = \int_{A^*} L(\theta_0) - \int_A L(\theta_0) = (1 - \alpha) - (1 - \alpha) = 0$$

and we have

$$\int_{A^*} L(\theta_1) - \int_A L(\theta_1) \geq 0$$

which establishes the desired result. The inequality is actually strict so long as $A^* \cap \bar{A} \neq \phi$; that is so long as $A^* \neq A$.

This result then gives us a concrete way of finding the "best" possible acceptance region, for tests of size α, when testing a simple hypothesis versus a simple alternative. Let us now examine some examples of its use.

Example 5.2.1

Assume X is known to be a normal random variable with $\sigma = 1$ and we want to test H_0: $\mu = 0$ versus H_1: $\mu = 1$ based on a sample of n

observations of X. Then

$$L(0) = \left(\frac{1}{2\pi}\right)^{n/2} e^{-\frac{1}{2}\Sigma x_i^2}$$

$$L(1) = \left(\frac{1}{2\pi}\right)^{n/2} e^{-\frac{1}{2}\Sigma (x_i-1)^2}$$

and the Neyman-Pearson acceptance region is

$$A: \quad \frac{L(1)}{L(0)} = \frac{e^{-\frac{1}{2}\Sigma(x_i-1)^2}}{e^{-\frac{1}{2}\Sigma x_i^2}} = e^{-\frac{1}{2}\{\Sigma(x_i-1)^2 - \Sigma x_i^2\}} < k.$$

Since $\sum (x_i - 1)^2 - \sum x_i^2 = -2n\bar{x} + n$, this region is equivalent to

$$A: \quad \bar{x} < \frac{1}{n} \log k + \frac{1}{2} = c.$$

Thus, to have probability of type I error equal to α we should choose c so that

$$P(\bar{X} < c \mid \mu = 0) = 1 - \alpha;$$

since, given $\mu = 0$ we know that $\bar{X}$ is $N(0, 1/\sqrt{n})$, we should clearly take $c = (1/\sqrt{n})z_{1-\alpha}$. Among all tests of this hypothesis, of size α, none has a smaller probability of type II error. Note that this minimum probability of type II error is

$$\beta = P\left(\bar{X} < \frac{1}{\sqrt{n}} z_{1-\alpha} \mid \mu = 1\right) = N_Z(z_{1-\alpha} - \sqrt{n})$$

where N_Z is the standard normal distribution function. Basing the test on the value that $\bar{X}$ has, in this example, certainly is the result we should expect, when we recall that $\sum X_i$ is the sufficient statistic for μ. Note that as $n \to \infty$, $\beta \to 0$, as we should expect it to. α, of course, remains fixed at its given value. Note that in this example if we choose $c = 1/2$, no matter what the value for n, the probabilities of type I and type II error are equal, and both would approach 0 as $n \to \infty$.

The following example applies the Neyman-Pearson lemma to a hypothesis about a vector valued parameter.

Example 5.2.2

It is possible that a female fly mated with either of two different genotype male flies. In either case, her offspring will fall into three distinct categories. If she mated with a male of genotype 1, the expected proportions of the three types of offspring are r_1, r_2, r_3, whereas, if she mated with a male of genotype

2 these expected proportions are q_1, q_2, q_3. Given that she produces n offspring from this mating and that their categories are determined like independent multinomial trials with parameters p_1, p_2, p_3, we then would be interested in testing the simple hypothesis

$$H_0: \quad \begin{pmatrix} p_1 \\ p_2 \\ p_3 \end{pmatrix} = \begin{pmatrix} r_1 \\ r_2 \\ r_3 \end{pmatrix} \text{ versus the simple alternative } H_1: \quad \begin{pmatrix} p_1 \\ p_2 \\ p_3 \end{pmatrix} = \begin{pmatrix} q_1 \\ q_2 \\ q_3 \end{pmatrix},$$

given the observed distribution among the categories. Applying the Neyman-Pearson lemma, the best acceptance region is

$$A: \quad \frac{L(q_1, q_2, q_3)}{L(r_1, r_2, r_3)} = \frac{(q_1)^{x_1}(q_2)^{x_2}(q_3)^{x_3}}{(r_1)^{x_1}(r_2)^{x_2}(r_3)^{x_3}} < k$$

where x_1, x_2, x_3 are the observed numbers of offspring in the three categories. This region is equivalent to

$$A: \quad \sum_{i=1}^{3} x_i \log\left(\frac{q_i}{r_i}\right) < \log k$$

so the best acceptance region reduces to looking at a weighted sum of the numbers in the three categories and accepting or rejecting H_0, depending on the value of this weighted sum.

It will be recalled from Theorem 3.3.2 that if we have a sufficient statistic $g(X_1, X_2, \ldots, X_n)$ for a parameter θ, then the joint density function for the sample, and thus the likelihood function, can be written

$$L(\theta) = G(g(x_1, x_2, \ldots, x_n))H(x_1, x_2, \ldots, x_n)$$

where H is independent of θ. Notice then that the best acceptance region

$$A: \quad \frac{L(\theta_1)}{L(\theta_0)} = \frac{G_1(g(x_1, \ldots, x_n))}{G_0(g(x_1, \ldots, x_n))} < k$$

gives an inequality involving only the observed value of the sufficent statistic $g(x_1, \ldots, x_n)$ (and θ_1, θ_0 and k), since $H(x_1, \ldots, x_n)$ is the same for any value of θ and cancels off in the ratio. Thus the Neyman-Pearson best acceptance regions will always be expressible in terms of the sufficient statistic for the parameter and there is no need to consider any other aspects of the sample values. When a test involves only the observed value of such a single statistic, the statistic involved is frequently called the *test statistic* for the given hypothesis.

EXERCISE 5.2

1. Prove Theorem 5.2.1 for a discrete random variable.

2. In Example 5.2.1, find c and n so that $\alpha = .01$, $\beta = .05$.

3. Assume X is a Poisson random variable. Given a random sample of n observations, what is the best acceptance region for testing H_0: $\lambda = 1$ versus H_1: $\lambda = 2$? How does it change if we have H_1: $\lambda = 1/2$? Assuming n is large, use the standard normal table to approximate the value for k in these two cases.

4. The probability that a given machine produces a defective item is p and the quality of the items produced varies independently from one to another. Given a random sample of $n = 20$ items produced by the machine, what is the form of the best acceptance region for testing H_0: $p = .05$ versus H_1: $p = .1$? What are the possible values of α for this case and the corresponding values for β, the probability of a type II error?

5. Assume that X is normal with known mean μ. What is the test statistic for the best test of H_0: $\sigma^2 = 1$ versus H_1: $\sigma^2 = 2$? Show how the χ^2 tables can be used to determine α and β.

6. X is a normal random variable. Given a random sample of n values of X, what is the form of the best acceptance region for testing H_0: $\mu = \mu_0$, $\sigma = \sigma_0$ versus H_1: $\mu = \mu_1$, $\sigma = \sigma_1$ where

(a) $\mu_1 > \mu_0$, $\sigma_1 > \sigma_0$
(b) $\mu_1 > \mu_0$, $\sigma_1 < \sigma_0$
(c) $\mu_1 < \mu_0$, $\sigma_1 > \sigma_0$
(d) $\mu_1 < \mu_0$, $\sigma_1 < \sigma_0$.

7. Assume $\mathbf{X}$ is a $p \times 1$ multivariate normal vector and $\mathbf{\Sigma} = \mathbf{I}$. What is the form of the best critical region for testing H_0: $\boldsymbol{\mu} = \mathbf{0}$ versus H_1: $\boldsymbol{\mu} = \boldsymbol{\mu}_1$ where $\boldsymbol{\mu}_1 \neq \mathbf{0}$? If $\boldsymbol{\mu} = \mathbf{0}$, what is the best critical region for testing H_0: $\mathbf{\Sigma} = \mathbf{I}$ versus H_1: $\mathbf{\Sigma} = 2\mathbf{I}$?

8. Given that the time to failure of a vacuum tube is an exponential random variable with parameter λ, let $R(t)$ be the reliability of one of these tubes for a fixed length of time t. Given a random sample of n lifetimes of this type of tube, what is the form of the best critical region for testing H_0: $R(t) = .99$ versus H_1: $R(t) = .999$? With $n = 10$, explicitly determine the acceptance region for $\alpha = .05$, $t = 1000$. (Hint: Use the χ^2 tables.) What is the power of this test?

9. Assume the same situation as in problem 8 except the n tubes are only tested for a fixed number T_0 of hours (see Example 3.2.6). What is the form of the best critical region for testing H_0: $R(T_0) = .99$ versus H_1: $R(T_0) = .999$?

10. X is a gamma random variable with $\lambda = 1$. Given a sample of n values of X, what is the form of the best acceptance region for testing H_0: $r = 5$ versus H_1: $r = 7$?

11. X is a gamma random variable with known parameter r. Given a sample of n values of X, what is the form of the best acceptance region for testing H_0: $\lambda = 1$ versus H_1: $\lambda = 2$? Is this same region the best for H_0: $\lambda = 1$ versus H_1: $\lambda = 3$? for H_1: $\lambda = \lambda_1$, where λ_1 is any value exceeding 1?

12. The Neyman-Pearson lemma can also be used to test hypotheses about the distribution of a random variable, provided both H_0 and H_1 are simple. Suppose, for example, we wanted to test the hypothesis H_0 that X is a beta random variable, $\alpha = 3/2$, $\beta = 3/2$, versus the hypothesis H_1 that X is a uniform random variable on the interval (0, 1). Based on a sample of n observations of X, what is the form of the best acceptance region for this hypothesis?

13. Given that X is a geometric random variable with parameter p, what is the form of the best acceptance region for testing H_0: $p = 1/2$ versus H_1: $p = 3/4$?

14. Given X is a negative binomial random variable with parameter r known, what is the form of the best acceptance region for testing H_0: $p = 1/2$ versus H_1: $p = 3/4$?

15. Given X is a negative binomial random variable with $p = 1/2$, what is the form of the best acceptance region for testing H_0: $r = 1$ versus H_1: $r = 2$?

16. Assume we have two independent random samples of size n, one each from two normal populations; both populations have $\sigma^2 = 1$. What is the best acceptance region for testing H_0: $\mu_1 = \mu_2 = 0$ versus H_1: $\mu_1 = 0$, $\mu_2 = 1$?

17. In Example 5.2.2, find the probability generating function for $\sum_{i=1}^{3} X_i \log[q_i/r_i]$ and show how this could be used to assign sample outcomes to the acceptance region A.

5.3 Composite Hypotheses; Generalized Likelihood Ratio

In Section 5.2 we examined simple hypotheses versus simple alternatives, and found that the Neyman-Pearson lemma provides the best test in such cases. In most applications H_0 or H_1 or both are composite. As was noted in Section 5.1, if H_1 is composite then the power of the test is a function, taking on a (possibly) different value for each of the alternative values given. Thus, in considering the alternative hypothesis, in such cases, we do not have a single value for the power function to maximize. Still, in certain situations, the acceptance region given by the Neyman-Pearson lemma is the best.

A frequent hypothesis of interest is that a parameter is equal to a specified value, versus the alternative that it is larger (or smaller) than the specified value. These are called *one-sided alternatives* (or one-sided hypotheses).

For example, a new synthetic rubber might be used in making automobile tires. If we know from past experience that the usual type of rubber gives an average "lifetime" of 25,000 miles, in standard usage, we might be interested in testing H_0: $\mu = 25{,}000$ versus H_1: $\mu > 25{,}000$, where μ is the average lifetime of tires made with the new synthetic rubber. If we accepted H_0 there would not seem to be any reason to switch to the new type of rubber, whereas, if we rejected H_0 we certainly would want to consider the use of the new synthetic in place of the standard. H_1 is called a one-sided alternative hypothesis. Frequently, when testing a simple H_0 versus a one-sided composite H_1, the acceptance region given by the Neyman-Pearson fundamental lemma is the same for each of the simple hypotheses specified by H_1; thus, for each simple hypothesis we have the most powerful test. We then say the test is *uniformly most powerful.* The following example is of this type.

Example 5.3.1

Assume that the length of life of an automobile tire is a normal random variable with unknown mean μ and standard deviation $\sigma = 3000$. To test H_0: $\mu = 25{,}000$ versus H_1: $\mu = \mu_1$, where $\mu_1 > 25{,}000$ we have seen that the Neyman-Pearson acceptance region is $\bar{X} < c$, where $P(\bar{X} < c \mid \mu = 25{,}000) = \alpha$ (see Example 5.2.1); no matter what the given value of μ_1 may be, the definition of A is the same. Thus if $\mu_1 = 26{,}000$ or $\mu_1 = 30{,}000$, or any value greater than 25,000, we should accept H_0 if $\bar{X} < c$ and we know there is no other test of size α which has greater power against H_1: $\mu = \mu_1$ where $\mu_1 > 25{,}000$. Thus, this test is *uniformly most powerful.* If the alternative had been H_1: $\mu < 25{,}000$ then the acceptance region would have been $\bar{X} > c$, where now c is chosen so that $P(\bar{X} > c \mid \mu = 25{,}000) = \alpha$; this again would be uniformly most powerful for the given one-sided alternative. Note immediately that if we wanted to test H_0: $\mu = 25{,}000$ versus H_1: $\mu \neq 25{,}000$ (H_1 is called a *two-sided alternative*) we would want to reject H_0 if either $\bar{X}$ were much larger or much smaller than 25,000; it would seem reasonable to define A by $P(-k < \bar{X} - 25{,}000 < k \mid \mu = 25{,}000) = 1 - \alpha$ and this test is not uniformly most powerful, since, for any α, the tests just described for the one-sided alternatives are uniformly most powerful and neither is identical to this one. Thus, for this case of a two-sided alternative hypothesis it is impossible to find a uniformly most powerful test. Figure 5.3.1 gives a graph of these three power functions. The curve labelled 1 is the power function for the test against H_1: $\mu > 25{,}000$, the curve labelled 2 is the test against H_1: $\mu < 25{,}000$ and the curve labelled 3 is the suggested test against H_1: $\mu \neq 25{,}000$. Clearly the suggested test is better than the ones for the one-sided alternatives, even though it is not uniformly most powerful, because it does provide protection

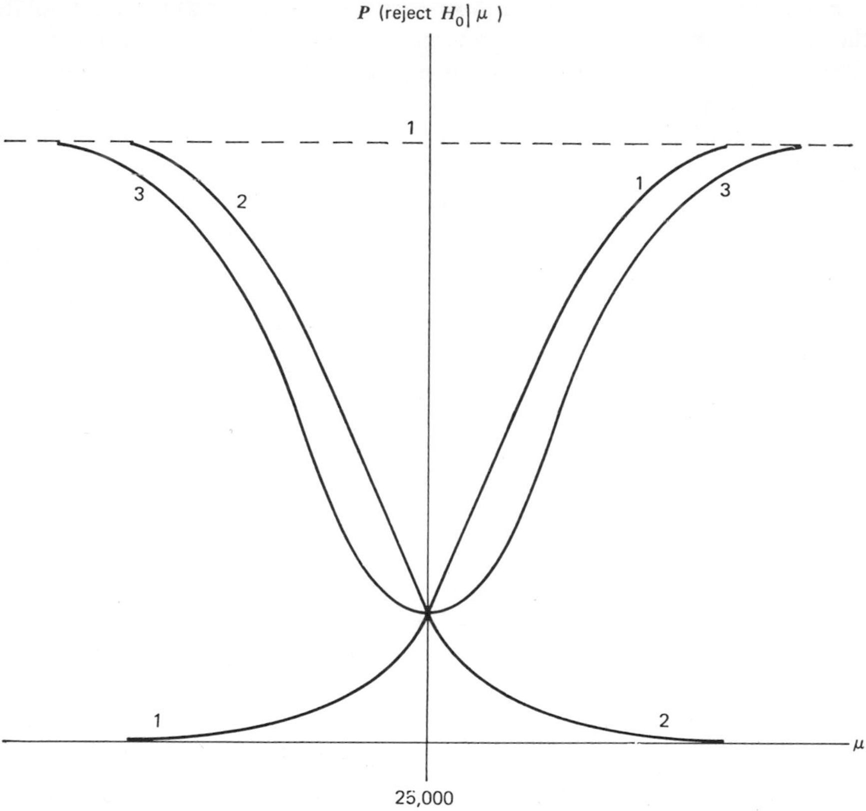

FIGURE 5.3.1. Comparison of power functions (see Example 5.3.1).

against accepting H_0 if μ lies on *either* side of 25,000; each of the one-sided tests provides the best possible protection against alternatives on its side and virtually none against alternatives on the other side of the value specified by H_0.

In many cases the Neyman-Pearson lemma yields a uniformly most powerful test of a simple hypothesis versus a one-sided alternative (see the exercises below for further examples). In fact, there is no need to restrict H_0 to being simple in many cases. In the above example, we could test H_0: $\mu \leq 25{,}000$ versus H_1: $\mu > 25{,}000$ letting A be $\bar{X} < c$, as done in that example; in this case we have simply extended the domain of H_0 to include the values of $\mu < 25{,}000$ and, for any $\mu > 25{,}000$, we still have the most powerful test. Note in this case that the probability of a type I error is itself a function and achieves its maximum value at $\mu = 25{,}000$ (see Figure 5.3.1).

For more general situations than testing a simple H_0 versus a simple H_1, the *(generalized) likelihood ratio test criterion* usually gives a good test. This procedure is as follows: let ω_0 be the set of values for the parameters specified by H_0, and let ω_1 be the set of values for the parameters specified by H_1. Define $\Omega = \omega_0 \cup \omega_1$; in many cases Ω will be the full *parameter space*, the set of all possible parameter values such that the density of the random variable sampled is a legitimate density function. Then, to construct a critical region for testing H_0 versus H_1 we compute the maximum value of the likelihood function, $L(\hat{\omega}_0)$, where the parameters are restricted to the values specified by ω_0; we also compute the maximum value of the likelihood function, $L(\hat{\Omega})$, where the parameters are restricted to values specified by Ω. Note that if Ω is the full parameter space then necessarily $L(\hat{\Omega})$ occurs for the parameters set equal to their maximum likelihood estimates. Since $\omega_0 \subset (\omega_0 \cup \omega_1) = \Omega$, we must have $L(\hat{\omega}_0) \leq L(\hat{\Omega})$ (the relative maximum, restricting the parameters to ω_0, cannot exceed the overall maximum for all parameter values in $\Omega = \omega_0 \cup \omega_1$). Thus, if we define $\lambda = L(\hat{\omega}_0)/L(\hat{\Omega})$, then $\lambda \leq 1$. If λ is close to one, then the value of $L(\hat{\omega}_0)$ is close to the overall maximum and the sample results are more or less consistent with H_0 being true. On the other hand, if λ is close to zero, then $L(\hat{\omega}_0)$ is much smaller than the overall maximum and the sample results do not seem particularly consistent with H_0. Thus, the likelihood ratio test criterion suggests we reject H_0 if $\lambda \leq \lambda_0$, where λ_0 is some positive fraction. The value of λ_0 is chosen so that the probability (a priori) of rejecting H_0, when it is true, is no bigger than α.

This procedure is entirely intuitive, albeit quite rational. You are asked in Exercise 3 below to show that this procedure is equivalent to that described in the Neyman-Pearson lemma, when both H_0 and H_1 are simple (thus when ω_0 and ω_1 each have only one element). The following examples should aid in understanding this criterion.

Example 5.3.2

Let us return to the two-sided alternative hypothesis mentioned in Example 5.3.1 and see what the likelihood ratio test is for this case. We assume X is a normal random variable with unknown mean μ and unknown variance σ^2 (note that this is more realistic than the assumption made in 5.3.1 that σ was known). Based on a sample of n observations of X, we want to test H_0: $\mu = 25{,}000$ versus H_1: $\mu \neq 25{,}000$. (H_0 is composite, since σ^2 is unknown and its value is unspecified.) Thus we have

$$\omega_0 = \{(25{,}000, \sigma^2): \quad \sigma^2 > 0\}$$

$$\omega_1 = \{(\mu, \sigma^2): \quad -\infty < \mu < \infty, \quad \mu \neq 25{,}000, \quad \sigma^2 > 0\};$$

and

$$\Omega = \omega_0 \cup \omega_1 = \{(\mu, \sigma^2): \quad -\infty < \mu < \infty, \quad \sigma^2 > 0\}$$

is the whole parameter space. The likelihood function for the sample is

$$L(\mu, \sigma^2) = \left(\frac{1}{2\pi\sigma^2}\right)^{n/2} e^{-(1/2\sigma^2)\sum_{i=1}^{n}(x_i-\mu)^2};$$

in Ω, the values of μ and σ^2 which maximize L are, from Chapter 3, $\hat{\mu} = \bar{x}$, $\hat{\sigma}^2 = (1/n)\sum_{i=1}^{n}(x_i - \bar{x})^2$. Substituting these values for μ and σ^2 in L we have

$$L(\hat{\Omega}) = \left(\frac{n}{2\pi\sum(x_i - \bar{x})^2}\right)^{n/2} e^{-(n/2)}.$$

In ω_0, $\mu = 25{,}000$ and thus we need only to maximize

$$L(25{,}000, \sigma^2) = \left(\frac{1}{2\pi\sigma^2}\right)^{n/2} e^{-(1/2\sigma^2)\Sigma(x_i-25{,}000)^2},$$

with respect to σ^2, to find $L(\hat{\omega}_0)$; taking the log of L, differentiating with respect to σ^2, and setting this derivative equal to zero easily gives $\hat{\sigma}_0^2 = (1/n)\sum(x_i - 25{,}000)^2$ as the value which maximizes $L(25{,}000, \sigma^2)$. Thus,

$$L(\hat{\omega}_0) = \left(\frac{n}{2\pi\sum(x_i - 25{,}000)^2}\right)^{n/2} e^{-n/2}$$

and the ratio $\lambda = [L(\hat{\omega}_0)]/[L(\hat{\Omega})]$ is

$$\lambda = \left(\frac{\sum(x_i - \bar{x})^2}{\sum(x_i - 25{,}000)^2}\right)^{n/2};$$

it is easily seen that

$$\sum(x_i - 25{,}000)^2 = \sum(x_i - \bar{x})^2 + n(\bar{x} - 25{,}000)^2$$

and, taking the $2/n$ root of λ and dividing numerator and denominator by $\sum(x_i - \bar{x})^2$ we then have

$$\lambda^{2/n} = \frac{1}{1 + \dfrac{n(\bar{x} - 25{,}000)^2}{\sum(x_i - \bar{x})^2}} = \frac{1}{1 + \dfrac{t^2}{n-1}}$$

where

$$t = \frac{\bar{x} - 25{,}000}{s/\sqrt{n}},$$

$$s^2 = \frac{1}{n-1}\sum(x_i - \bar{x})^2,$$

the sample variance. Then the likelihood ratio critical region is

$$\lambda \le \lambda_0$$

or

$$\lambda^{2/n} \le \lambda_0^{2/n};$$

thus for points in the critical region we want

$$\frac{1}{1 + \dfrac{t^2}{n-1}} \le \lambda_0^{2/n}$$

or

$$|t| \ge \sqrt{n-1}(\lambda_0^{-(2/n)} - 1)^{1/2}.$$

Thus the likelihood ratio test criterion is equivalent to rejecting H_0 if the observed variable $|t|$, defined above, gets too large. As we saw in Section 2.3, $T = (\bar{X} - 25{,}000)/(S/\sqrt{n})$ has the t distribution with $n - 1$ degrees of freedom, if X is normal with mean 25,000 and variance σ^2. Thus, if we reject H_0 if $|T| \ge t_{n-1}[1 - \alpha/2]$, we have probability of type I error equal to α and the likelihood ratio test of H_0: $\mu = 25{,}000$ utilizes the t distribution.

Example 5.3.3

In some problems it is important that the variance of the distribution be well controlled. For example, in manufacturing a cylindrical machine part which must fit inside another, the diameter of the cylinder must not vary too much or we will have a lot of parts which will either not fit at all inside their mates or will fit too loosely, even if the machine setting produces the right average diameter. Assume that for a given setting on the machine, the diameter of the cylinders produced is a normal random variable X with mean μ and variance σ^2. Given a sample of n diameters produced with this setting, then, we might want to test H_0: $\sigma^2 \le \sigma_0^2$ versus H_1: $\sigma^2 > \sigma_0^2$. Then

$$\omega_0 = \{(\mu, \sigma^2): \quad -\infty < \mu < \infty, \sigma^2 \le \sigma_0^2\},$$
$$\omega_1 = \{(\mu, \sigma^2): \quad -\infty < \mu < \infty, \sigma^2 > \sigma_0^2\}$$

and $\Omega = \omega_0 \cup \omega_1$ is again the full parameter space. In Ω the maximizing values are again $\hat{\mu} = \bar{x}$, $\hat{\sigma}^2 = (1/n) \sum (x_i - \bar{x})^2$. In ω_0, μ can take on any real value but σ^2 is restricted to values no bigger than σ_0^2. If $\hat{\sigma}^2 = (1/n) \sum (x_i - \bar{x})^2 \le \sigma_0^2$, then the maximizing value for σ^2 is $\hat{\sigma}^2$, and we would obviously have $L(\hat{\omega}_0) = L(\hat{\Omega})$, thus $\lambda = 1$, and any sample outcomes with $\hat{\sigma}^2 \le \sigma_0^2$ are automatically in A, the acceptance region. On the other hand, if $\hat{\sigma}^2 > \sigma_0^2$, then the maximizing value of $\sigma^2 \in \omega_0$ is σ_0^2. Thus, for

$\hat{\sigma}^2 > \sigma_0^2$ the likelihood ratio is

$$\lambda = \frac{\left(\dfrac{1}{2\pi\sigma_0^2}\right)^{n/2} e^{-\sum(x_i-\bar{x})^2/2\sigma_0^2}}{\left(\dfrac{n}{2\pi\sum(x_i-\bar{x})^2}\right)^{n/2} e^{-(n/2)}}$$

$$= \left(\frac{\hat{\sigma}^2}{\sigma_0^2}\right)^{n/2} e^{-n\hat{\sigma}^2/2\sigma_0^2+n/2}, \qquad \text{for } \frac{\hat{\sigma}^2}{\sigma_0^2} > 1.$$

The graph of λ versus $\hat{\sigma}^2/\sigma_0^2$, then, is easily seen to be as given in Figure 5.3.2. Since λ is a function of $\hat{\sigma}^2/\sigma_0^2$, the critical region $\lambda \leq \lambda_0$ is equivalent to $\hat{\sigma}^2/\sigma_0^2 \geq k$. As we saw in Section 2.3, $\sum (X_i - \bar{X})^2/\sigma_0^2$ is a χ^2 random variable with $n - 1$ degrees of freedom if $\sigma^2 = \sigma_0^2$. From the χ^2 tables we can find $\chi^2_{n-1}(1 - \alpha)$ such that

$$P\left(\frac{\sum (X_i - \bar{X})^2}{\sigma_0^2} \geq \chi^2_{n-1}(1 - \alpha)\right) = \alpha$$

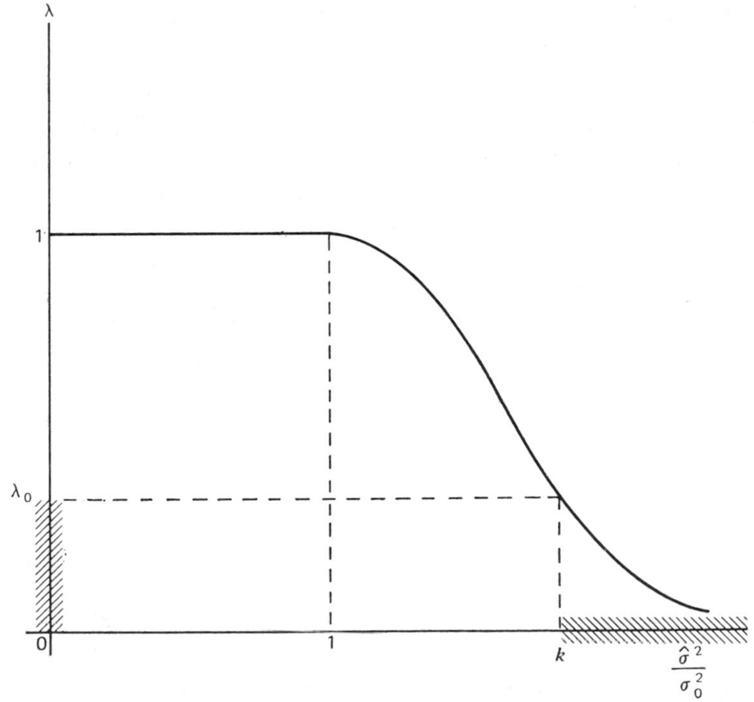

FIGURE 5.3.2. Graph of the likelihood ratio as a function of $u = \hat{\sigma}^2/\sigma_0^2$ (see Example 5.3.3).

and if we choose k above to be $(1/n)\chi^2_{n-1}(1 - \alpha)$, the maximum value of the probability of type I error is α.

Perhaps the reader has noticed that in both Examples 5.3.2 and 5.3.3, the likelihood ratio test criterion has led to acceptance regions which are intimately related to confidence intervals for the corresponding parameters. In Example 5.3.2 the acceptance region A is

$$\left|\frac{\bar{X} - \mu_0}{S/\sqrt{n}}\right| < t_{n-1}\left(1 - \frac{\alpha}{2}\right)$$

which is equivalent to

$$\bar{X} - \frac{S}{\sqrt{n}}\, t_{n-1}\left(1 - \frac{\alpha}{2}\right) < \mu_0 < \bar{X} + \frac{S}{\sqrt{n}}\, t_{n-1}\left(1 - \frac{\alpha}{2}\right)$$

(where $\mu_0 = 25{,}000$ in the example); the endpoints of this interval are exactly the confidence interval for μ which we discussed in Section 4.2. Thus, the likelihood ratio test criterion, in this example, is equivalent to taking the sample, computing a $100(1 - \alpha)\%$ confidence interval for μ, and then if the interval includes the hypothesized value μ_0, we accept H_0: $\mu = \mu_0$; if it does not include μ_0 then we reject H_0. The size of this test is α. The reader is asked to verify in Exercise 7 below that the test for σ^2 in Example 5.3.3 can also be looked at in this way.

This relationship between tests and confidence intervals can be seen to hold true in general. If (L_1, L_2) is a $100(1 - \alpha)\%$ confidence interval for an unknown parameter θ of a continuous distribution, then, by definition,

$$P(L_1 \leq \theta \leq L_2) \geq 1 - \alpha.$$

Thus, if in fact $\theta = \theta_0$, (L_1, L_2) will include θ_0 with probability at least $1 - \alpha$; if we reject H_0: $\theta = \theta_0$ for any sample in which the observed interval does not include θ_0, then we have probability at most α of incorrectly rejecting H_0. Thus in any case in which we are able to derive a confidence interval for an unknown parameter, we also have an associated test of the hypothesis that the parameter takes on any specified value. And, conversely, if we are able to find a test of size α of H_0: $\theta = \theta_0$ versus H_1: $\theta \neq \theta_0$, the acceptance region for the test forms a $100(1 - \alpha)\%$ confidence interval for θ. (If θ is a vector of parameters then we can obviously also relate confidence regions for θ to tests that each parameter in the vector takes on a specified value.)

In the two examples just concluded we did not discuss the power of the tests derived. In Example 5.3.2 the acceptance region was $|T| \leq t_{n-1}(1 - \alpha/2)$ where $T = (\bar{X} - \mu_0)/(S/\sqrt{n})$. To compute the power of this test we want to know the value of $P(|T| \leq t_{n-1}[1 - \alpha/2] \mid \mu = \mu_1)$, where μ_1 is any possible value of μ not equal to μ_0; as μ_1 varies over the possible alternative values specified by H_1 we have the full power function. If in fact $\mu = \mu_1$, then $(\bar{X} - \mu_0)/(S/\sqrt{n})$ has the noncentral t distribution with $f = n - 1$ degrees of freedom and noncentrality parameter $\delta = [(\mu_1 - \mu_0)\sqrt{n-1}]/\sigma$ (see Section 2.3); thus

$$P\left(\frac{|\bar{X} - \mu_0|}{S/\sqrt{n}} \leq t_{n-1}\left(1 - \frac{\alpha}{2}\right) \middle| \mu = \mu_1\right)$$

can be computed from tables of the noncentral t distribution (see Lieberman and Resnikoff).

In Example 5.3.3, for testing H_0: $\sigma^2 \leq \sigma_0^2$ versus H_1: $\sigma^2 > \sigma_0^2$, the acceptance region was $\hat{\sigma}^2/\sigma_0^2 < k$. If in fact $\sigma^2 = \sigma_1^2 > \sigma_0^2$, then $[\sum (X_i - \bar{X})^2]/\sigma_1^2$ is χ^2_{n-1} and $\hat{\sigma}^2/\sigma_0^2 < k$ is equivalent to $\sum (X_i - \bar{X})^2/\sigma_1^2 < (\sigma_0^2/\sigma_1^2)kn$. Thus the power for any given alternative σ_1^2 is equal to $P[\chi^2_{n-1} \geq (\sigma_0^2/\sigma_1^2)kn]$ which can be computed, at least approximately, from a table of the χ^2_{n-1} distribution.

EXERCISE 5.3

1. Is the acceptance region derived in Exercise 5.2.3 the uniformly most powerful test of H_0: $\lambda \leq 1$ versus H_1: $\lambda > 1$?

2. Is the acceptance region derived in Exercise 5.2.4 the uniformly most powerful test of H_0: $p \leq .05$ versus H_1: $p > .05$?

3. Show that if H_0 and H_1 are both simple, and thus ω_0 and ω_1 are both one-element sets, the likelihood ratio test criterion critical region is the same as the Neyman-Pearson critical region.

4. Derive the likelihood ratio test criterion test of H_0: $\sigma = \sigma_0$ versus H_1: $\sigma = \sigma_1$, given a random sample of n observations of a normal random variable with mean μ and variance σ^2, with μ unknown. Is the criterion the same if μ is known?

5. What is the generalized likelihood ratio test of H_0: $\mu = \mu_0$ versus H_1: $\mu \neq \mu_0$, for a normal random variable with known variance?

6. In Example 5.3.3 the statement was made that $L(\hat{\omega}_0) = L(\hat{\Omega})$ if $\hat{\sigma}^2 \leq \sigma_0^2$. Provide the details of this argument. (Hint: Look at the slope of $L(\mu, \sigma^2)$ with μ fixed.)

7. Show that the test of H_0: $\sigma^2 \leq \sigma_0^2$ versus H_1: $\sigma^2 > \sigma_0^2$, in Example 5.3.3, is equivalent to computing a one-sided confidence region for σ^2 and accepting H_0 only if the interval includes σ_0^2.

8. In Theorem 4.2.5, we saw a confidence region for μ and σ for a normal random variable. If we select a sample, compute this region and accept H_0: $\mu = \mu_0$, $\sigma = \sigma_0$, reject H_1: $\mu \neq \mu_0$ or $\sigma \neq \sigma_0$ or both, if and only if the region includes μ_0, σ_0, are we using the likelihood ratio test?

9. The time to failure of a certain type of electron tube is an exponential random variable with unknown parameter λ. N tubes are put on test and the test is concluded when k (fixed in advance) have failed. What is the likelihood ratio test criterion of H_0: $\lambda \geq \lambda_0$ versus H_1: $\lambda < \lambda_0$?

10. (Refer to Exercise 3.2.31). Assume $3N$ flies are in a box, N females, all of the same genotype, as well as $2N$ males, 1/2 of which are of genotype 1 and the other half of which are of genotype 2. Again, two different distributions of descendants (of two different types) are known to occur, depending on whether a female mated with a male of type 1 or type 2. The probability any given female prefers to mate with a male of type 1 is p. What is the likelihood ratio test of H_0: $p = 1/2$ versus H_1: $p \neq 1/2$?

11. Given two independent random samples, each of size n, from two different normal populations, what is the likelihood ratio test of H_0: $\mu_1 = \mu_2$ versus H_1: $\mu_1 \neq \mu_2$, assuming $\sigma_1^2 = \sigma_2^2$? What is the test if we do not assume $\sigma_1^2 = \sigma_2^2$?

12. (*Paired t test*). Assume a random sample of size n of a bivariate normal vector $\mathbf{X}$ with

$$\boldsymbol{\mu} = \begin{pmatrix} \mu_1 \\ \mu_2 \end{pmatrix}, \qquad \boldsymbol{\Sigma} = \begin{pmatrix} \sigma_{11} & \sigma_{12} \\ \sigma_{21} & \sigma_{22} \end{pmatrix}.$$

Note that if we define $\mathbf{c}' = (1, -1)$, then we have a sample of n observations of $Y = \mathbf{c}'\mathbf{X}$. What is the mean and variance of Y? How would we test that the mean of Y is 0, versus the alternative that it is not? Show that this test is equivalent to H_0: $\mu_1 = \mu_2$ versus H_1: $\mu_1 \neq \mu_2$, and thus, even though the two variances are not equal (and they may even be correlated), we can in this case get a simple test of $\mu_1 = \mu_2$; note, however, we have assumed the special situation of the observations from the two populations being naturally paired together.

5.4 Other Tests

We have already studied the Neyman-Pearson lemma, giving the "best" test of a simple hypothesis versus a simple alternative, and the generalized likelihood ratio test criterion which can be used for testing composite hypotheses. In this section we shall study a number of tests, including a "goodness of fit" test (which is used in testing the second type of hypothesis we discussed in Section 5.1, whether a random variable appears to have a specified distribution) all of which are based on a test statistic which, at least for large samples, has a χ^2 distribution. This type of test is not generally the same as those generated by the likelihood ratio test criterion.

As we know from the central limit theorem, $(X - np)/\sqrt{npq}$ is for large n approximately a standard normal random variable, where X is binomial with parameters n and p. Thus, for large n, $(X - np)^2/npq$ is approximately a χ^2 random variable with one degree of freedom, since it is (approximately) the square of a standard normal random variable. The algebraic identity

$$\frac{(X-np)^2}{np(1-p)} = \frac{(X-np)^2}{np} + \frac{(-X+np)^2}{n(1-p)}$$

$$= \frac{(X-np)^2}{np} + \frac{[(n-X)-n(1-p)]^2}{n(1-p)}$$

can be used then to establish that the sum of the two quantities $(X - np)^2/np$ and $[(n - X) - n(1 - p)]^2/n(1 - p)$ is approximately a χ^2 random variable with one degree of freedom. Notice that in observing our n repeated, independent Bernoulli trials, each is classified as being a success or failure; X is the total number of successes, in the n trials. If the probability of success on each trial is p, then the expected number of successes is np and the expected number of failures is $n(1 - p)$; thus, in taking our sample of size n, we have classified each observation as coming from one of two classes (success or failure); we can, knowing p, compute the expected number in each class, and if we square the difference between the number observed and the number expected, then divide by the number expected and sum over the two classes, the sum of the two terms is approximately a χ^2 random variable with one degree of freedom. A generalization of this result to more than two classes is very useful in testing a variety of hypotheses; we shall examine this generalization in this section, and see some of the hypotheses it is commonly used to test. The following theorem gives the generalization. We shall indicate the method of proof.

THEOREM 5.4.1. Assume $(X_1, X_2, \ldots, X_k)$ is the multinomial random vector with parameters $n, p_1, p_2, \ldots, p_k$. As n tends to infinity the asymptotic distribution of

$$V = \sum_{i=1}^{k} \frac{(X_i - np_i)^2}{np_i}$$

is χ^2 with $k - 1$ degrees of freedom.

Proof: Let us examine the case with $k = 3$ in some detail. Recall that if we have $k = 3$ different classes for each outcome, then actually we only have two random variables, since $X_3 = n - X_1 - X_2$. We saw in Section 1.4 that $\mathbf{X} = \begin{pmatrix} X_1 \\ X_2 \end{pmatrix}$ has mean vector $\boldsymbol{\mu} = \begin{pmatrix} np_1 \\ np_2 \end{pmatrix}$ and variance covariance matrix

$$\boldsymbol{\Sigma} = \begin{pmatrix} np_1(1-p_1) & -np_1p_2 \\ -np_1p_2 & np_2(1-p_2) \end{pmatrix}.$$

Then, for any n, if we define

$$\mathbf{Y} = \begin{pmatrix} Y_1 \\ Y_2 \end{pmatrix} = \begin{pmatrix} \dfrac{1}{\sqrt{np_1(1-p_1)}} & 0 \\ 0 & \dfrac{1}{\sqrt{np_2(1-p_2)}} \end{pmatrix} (\mathbf{X} - \boldsymbol{\mu})$$

it is easily seen that $\mathbf{Y}$ has mean vector $\mathbf{0}$ and covariance matrix

$$\boldsymbol{\Sigma}_{\mathbf{Y}} = \begin{pmatrix} 1 & -\sqrt{\dfrac{p_1 p_2}{(1-p_1)(1-p_2)}} \\ -\sqrt{\dfrac{p_1 p_2}{(1-p_1)(1-p_2)}} & 1 \end{pmatrix}.$$

It is shown in more advanced courses that as n tends to infinity the distribution of $\mathbf{Y}$ tends to that of a two-dimensional normal (this is the analog of the one-dimensional central limit theorem) with mean vector $\mathbf{0}$ and variance covariance matrix $\boldsymbol{\Sigma}_{\mathbf{Y}}$ given above (see Cramér (7) for details). As we saw in Exercise 2.3.15, if $\mathbf{Y}$ is p-variate normal with $\mathbf{0}$ mean and variance covariance matrix $\boldsymbol{\Sigma}_{\mathbf{Y}}$, then $\mathbf{Y}'\boldsymbol{\Sigma}_{\mathbf{Y}}^{-1}\mathbf{Y}$ is a χ^2 random variable with p degrees of freedom (p is 2 here). It is easily seen that

$$\boldsymbol{\Sigma}_{\mathbf{Y}}^{-1} = \frac{(1-p_1)(1-p_2)}{1-p_1-p_2} \begin{pmatrix} 1 & \sqrt{\dfrac{p_1 p_2}{(1-p_1)(1-p_2)}} \\ \sqrt{\dfrac{p_1 p_2}{(1-p_1)(1-p_2)}} & 1 \end{pmatrix}$$

and thus

$$\begin{aligned}
\mathbf{Y}'\boldsymbol{\Sigma}_{\mathbf{Y}}^{-1}\mathbf{Y} &= \frac{(1-p_1)(1-p_2)}{1-p_1-p_2} \sum_{i=1}^{2} Y_i^2 + 2\,\frac{\sqrt{p_1(1-p_1)p_2(1-p_2)}}{1-p_1-p_2}\, Y_1 Y_2 \\
&= \frac{(X_1 - np_1)^2(p_1+p_3)}{np_1p_3} + \frac{(X_2 - np_2)^2(p_2+p_3)}{np_2p_3} \\
&\quad + 2\,\frac{(X_1 - np_1)(X_2 - np_2)}{np_3} \\
&= \frac{(X_1 - np_1)^2}{np_1} + \frac{(X_2 - np_2)^2}{np_2} + \frac{[(X_1 - np_1) + (X_2 - np_2)]^2}{np_3} \\
&= \frac{(X_1 - np_1)^2}{np_1} + \frac{(X_2 - np_2)^2}{np_2} + \frac{(X_3 - np_3)^2}{np_3}
\end{aligned}$$

where $p_3 = 1 - p_1 - p_2$, $X_3 = n - X_1 - X_2$, and the result is established for the special case of $k = 3$. The argument is exactly the same for higher dimensional cases, with the result that $\sum_{i=1}^{k} (X_i - np_i)^2/(np_i)$ is asymptotically a χ^2 random variable with $k - 1$ degrees of freedom.

Let us now apply this result in testing a hypothesis about a multinomial random variable. It should be remarked that many authors recommend using this χ^2 approximation so long as the expected number in each class is at least $5(np_i \geq 5$, for all $i)$.

Example 5.4.1

Suppose that on the basis of a large number, n, of tosses, we want to test the hypothesis that a given die is fair. Thus, letting X_i be the number of times that face i occurs, $i = 1, 2, \ldots, 6$, the die is fair if and only if $p_i = 1/6$, $i = 1, 2, \ldots, 6$. Then, approximately,

$$V = \sum_{i=1}^{6} \frac{(X_i - (n_i/6))^2}{(n_i/6)}$$

is a χ^2 random variable with 5 degrees of freedom. If we reject the hypothesis if $v \geq \chi_5^2(1 - \alpha)$, where v is the observed value of V, our probability of type I error is approximately α, for $n \geq 30$ (so $np_i \geq 5$ for all i).

The result given in Theorem 5.4.1 can be used to test hypotheses that random variables have any specified distribution, discrete or continuous. Suppose, for example, X is a discrete random variable with probability function $p_X(x)$. Let $A_1, A_2, \ldots, A_k$ be any partition of the range of X; thus $A_i \cap A_j = \phi$ for $i \neq j$ and $\bigcup_{i=1}^{k} A_i = R_X$. Then, so long as $p_X(x)$ contains no unknown parameters we can compute the probability that the value of X will lie in A_j by

$$p_j = \sum_{x \in A_j} p_X(x), \qquad j = 1, 2, \ldots, k$$

and, of course, $\sum_{j=1}^{k} p_j = 1$. Then, if we have a random sample of n observations of X, each of the X_i will fall into exactly one of the A_j's, $i = 1, 2, \ldots, n$, $j = 1, 2, \ldots, k$. Thus, if we let Y_j be the number of sample values to fall in A_j, $j = 1, 2, \ldots, k$, then $Y_1, Y_2, \ldots, Y_k$ is the multinomial vector with parameters $n, p_1, p_2, \ldots, p_k$ defined above. To test the hypothesis that $p_X(x)$, used to compute $p_1, p_2, \ldots, p_k$, is the correct probability function for the population sampled, we can use the approximate χ^2 statistic given in Theorem 5.4.1. If we want to test that a population random variable X has a specified density function, $f_X(x)$, based on a sample of n observations, we proceed in exactly the same way, except the p_i's are defined by $\int_{A_i} f_X(x)\,dx$ rather than by summation. Notice that k, the number of sets in the partition, is arbitrary; it is recommended that $np_i \geq 5$, for all i, and thus we should consider using k no larger than $n/5$. The following examples should help to clarify this type of test.

Example 5.4.2
The number of autos declared abandoned and disposed of in a town of 50,000 people, in the 104 weeks of 1970–1971, is given in the following table:

Number of weeks	Number of autos
40	0
43	1
17	2
4	3 or more

Let us use the test just described to test the hypothesis that X, the number of cars declared abandoned and disposed of, per week, is a Poisson random variable with parameter $\lambda = 1$, with $\alpha = .05$. To two decimal places, the probabilities that $X = 0, 1, 2, 3$ or more, and the expected number of occurrences in 104 weeks, if the hypothesis is correct, are as follows:

x	$P(X = x)$	Expected number (104 weeks)
0	.37	38.5
1	.37	38.5
2	.18	18.7
3 or more	.08	8.3

Then, to test the hypothesis, we compute

$$\frac{(40 - 38.5)^2}{38.5} + \frac{(43 - 38.5)^2}{38.5} + \frac{(17 - 18.7)^2}{18.7} + \frac{(4 - 8.3)^2}{8.3} = 2.96;$$

we find $\chi_3^2(.95) = 7.81$ and, since $2.96 < 7.81$, we would accept the hypothesis that X is Poisson with parameter $\lambda = 1$.

Example 5.4.3
The time to failure of 1000 electron tubes of the same type are summarized in the following table:

Time to failure t	Number of tubes
$t \leq 150$	543
$150 < t \leq 300$	258
$300 < t \leq 450$	120
$450 < t \leq 600$	48
$600 < t \leq 750$	20
$750 < t$	11

Let us test the hypothesis that Y, the time to failure of a tube of this type, is an exponential random variable with parameter $\lambda = .005$, versus the alternative that it is not, with $\alpha = .1$ (either the parameter has a different value, or the distribution is not exponential, or both). The values of the p_i's are computed by integrating the density function over the appropriate range; thus

$$p_1 = \int_0^{150} .005e^{-.005t}\,dt = .528,$$

$p_2 = .249$, $p_3 = .118$, $p_4 = .056$, $p_5 = .026$, $p_6 = .023$ and the observed value of v is

$$\frac{(543 - 528)^2}{528} + \frac{(258 - 249)^2}{249} + \cdots + \frac{(11 - 23)^2}{23} = 9.99;$$

since $\chi_5^2(.90) = 9.24 < 9.99$, we would reject the hypothesis.

This approximate χ^2 test that a random variable has a given distribution is appropriate only when the hypothesized distribution is completely specified, parameters and all. However, a slight adjustment in the procedure, and the degrees of freedom used in the χ^2 distribution, enables us to test hypotheses that the distribution of the random variable is of a specified type, without giving explicit values to the parameters involved. The following theorem, which we shall not prove, gives the method of doing this. The interested reader is referred to Cramér (7) for a proof of this result.

THEOREM 5.4.2. Assume X has distribution function F_X, which satisfies certain regularity conditions and is indexed by parameters $\theta_1, \theta_2, \ldots, \theta_r$, and let $A_1, A_2, \ldots, A_k$, $k > r + 1$, be a partition of R_X. Let $\hat{\Theta}_1, \hat{\Theta}_2, \ldots, \hat{\Theta}_r$ be the maximum likelihood estimators of the parameters, based on a random sample of n observations of X and define $P_i = P(X \in A_i)$, $i = 1, 2, \ldots, k$, where the parameters have been replaced by their maximum likelihood estimators. As $n \to \infty$

$$V = \sum_{i=1}^{k} \frac{(Y_i - nP_i)^2}{nP_i}$$

is asymptotically a χ^2 random variable with $k - r - 1$ degrees of freedom, where Y_i is the number of times $X \in A_i$ in the sample of n observations.

Thus, if we only want to test that a random variable has a specified form of distribution, we estimate the parameters by maximum likelihood, from the sample itself, and then proceed to compute the expected numbers as we did before. The form of the test statistic is identical to the previous case, the only change being that we must delete one degree of freedom for every parameter estimated. Heuristically, we do this because the maximum likelihood estimators of the parameters can be shown to (asymptotically) be the same as those that minimize the value of this χ^2 statistic (*minimum* χ^2 *estimation*). The reduction in the number of degrees of freedom compensates for the fact that the parameters are given values that (essentially) minimize the observed χ^2 statistic. The following example illustrates this technique.

Example 5.4.4
The actual flight time for a commercial jetliner from San Francisco to New York, from the time the wheels leave the ground until they touch down again, is a random variable X. Making the flight repeatedly, with the same type of aircraft, in essentially identical flight conditions, might be expected to give rise to a normal distribution (actually we are referring to the conditional distribution of the time, given that no mishaps occur and that the flight is successfully terminated). Assume that the records of 80 such flights, in minutes, were as recorded below:

Flight time t	Number of flights
$t \leq 245$	10
$246 \leq t \leq 250$	22
$251 \leq t \leq 255$	25
$256 \leq t \leq 260$	18
$t \geq 261$	5

We are also given that the sum of all the flight times was 20,160 and that the sum of their squares was 5,082,320; then, the maximum likelihood estimates of μ and σ^2 are $\hat{\mu} = 20{,}160/80 = 252$ and $\hat{\sigma}^2 = (1/80)(5{,}082{,}320 - 80(252)^2) = 25$. Using these estimates we can then compute the expected numbers given below. (Since the normal distribution is continuous, these numbers are computed with 245.5, 250.5, etc., as the boundaries between the classes.)

Flight time t	Expected number
$t \leq 245$	7.7
$246 \leq t \leq 250$	22.8
$251 \leq t \leq 255$	30.1
$t \geq 256$	19.4

Note that the two highest classes have been combined so that each expected number will be at least 5. The observed value of the test statistic is

$$\frac{(10-7.7)^2}{7.7}+\frac{(22-22.8)^2}{22.8}+\frac{(25-30.1)^2}{30.1}+\frac{(23-19.4)^2}{19.4}=2.87.$$

Since $\chi_1^2(.95) = 3.841 > 2.87$, we would accept the hypothesis of normality with $\alpha = .05$.

The random variable used in testing the hypothesis in this example has historically been used occasionally to generate estimators of unknown parameters. Suppose we have a sample of size n of a random variable X, and that we have partitioned R_X into $A_1, A_2, \ldots, A_k$; then, if in fact X is normal with unknown mean μ and unknown variance σ^2, we know that

$$p_i = \int_{A_i} \frac{1}{\sigma\sqrt{2\pi}} e^{-(x-\mu)^2/2\sigma^2}\,dx$$

and thus $p_1, p_2, \ldots, p_k$ are functions of μ and σ^2. Since, for large n,

$$V = \sum_{i=1}^{k} \frac{(Y_i - np_i)^2}{np_i},$$

is approximately a χ^2 random variable, where Y_i is the number of X_j's belonging to A_i, we might use as estimators of μ and σ^2 those values which give V its smallest possible value; these would be the values of the parameters that give the best fitting normal distribution, in terms of this particular test statistic. This procedure is called *minimum χ^2 estimation.*

Using the chain rule for differentiation notice that we have

$$\frac{\partial V}{\partial \mu} = -2\sum_{i=1}^{k}\left(\frac{(Y_i - np_i)}{p_i} + \frac{1}{2}\frac{(Y_i - np_i)^2}{np_i^2}\right)\frac{\partial p_i}{\partial \mu}$$

with a similar equation for $\partial V/\partial\sigma^2$. Finding the solutions to $\partial V/\partial\mu = 0$, $\partial V/\partial\sigma^2 = 0$ then yields the minimum χ^2 estimators for μ and σ^2. Notice that this system of equations will frequently be unwieldy to solve, without the use of a computer. The *modified minimum χ^2 estimators* are the solutions to

$$\sum_{i=1}^{k}\frac{(Y_i - np_i)}{p_i}\frac{\partial p_i}{\partial \mu} = 0$$

$$\sum_{i=1}^{k}\frac{(Y_i - np_i)}{p_i}\frac{\partial p_i}{\partial \sigma^2} = 0,$$

a system which is generally easier to solve. Both methods yield estimators with reasonable large sample properties, but the maximum likelihood estimators are still preferable, in general.

A χ^2 statistic can also be used to test the independence of two methods of classification. For example, suppose every person that will cast a vote for the governor of California can be classified as being a republican, democrat, or independent, by registration; furthermore, the vote he casts will be for the republican candidate, for the democratic candidate, or for someone other than these two. Thus, every voter can be classified according to the two criteria: registration and person voted for, and thus will fall in exactly one cell of the following 3×3 table.

		Registration	
	Republican	Democrat	Independent
Republican			
Votes for Democrat			
Other			

If party affiliation has much meaning in California, we would not expect these two methods of classification to be independent: in fact, we would expect that the preponderance of voters would be classified on the main diagonal. To test the hypothesis of independence, based on a sample of voters, we proceed as follows: Each voter can be thought of as being a multinomial trial with 9 possible outcomes, represented by the 9 cells in the 3×3 table above. It is reasonable to assume that if p_{ij} is the proportion of voters in the population that fall in the ijth cell, $i = 1, 2, 3, j = 1, 2, 3$, then each voter selected at random has these probabilities of falling in the appropriate cells. Thus, a sample of n voters can be looked at as a sample of n multinomial trials with probabilities p_{ij}, $i = 1, 2, 3, j = 1, 2, 3$. Furthermore, let

$$\sum_{i=1}^{3} p_{ij} = p_{.j}, \qquad j = 1, 2, 3$$

be the marginal proportions in the population that are registered republican, democrat, or independent, respectively, and let

$$\sum_{j=1}^{3} p_{ij} = p_{i.}, \qquad i = 1, 2, 3$$

be the marginal proportions that will vote republican, democrat, or other, respectively. Then, if in fact voter registration is independent of the way in which the votes will be cast, $p_{ij} = p_{i.}p_{.j}$, $i = 1, 2, 3, j = 1, 2, 3$.

Let X_{ij}, $i = 1, 2, 3, j = 1, 2, 3$ be the number of voters in the sample to fall in cell ij and let

$$X_{i.} = \sum_{j=1}^{3} X_{ij}$$

$$X_{.j} = \sum_{i=1}^{3} X_{ij}$$

be the marginal totals. Under the assumption of independence, $\hat{P}_{i.} = X_{i.}/n$ and $\hat{P}_{.j} = X_{.j}/n$ are the maximum likelihood estimators for $p_{i.}$ and $p_{.j}$, respectively. Thus, to test H_0: $p_{ij} = p_{i.}p_{.j}$ we can use the statistic

$$V = \sum_{i=1}^{3} \sum_{j=1}^{3} \frac{(X_{ij} - n\hat{P}_{i.}\hat{P}_{.j})^2}{n\hat{P}_{i.}\hat{P}_{.j}} ;$$

for large n, V is approximately a χ^2 random variable with $9 - 1 - 2 - 2 = 4 = 2 \times 2$ degrees of freedom. (For a general $r \times c$ table, r rows and c columns, we have $rc - 1 - (r - 1) - (c - 1) = (r - 1)(c - 1)$ degrees of freedom, since we would have $rc - 1$ degrees of freedom if all parameters were known and we had to estimate $r - 1$ parameters for the marginal row probabilities and $c - 1$ parameters for the marginal column probabilities.) If $V \geq \chi_4^2(1 - \alpha)$, we reject H_0; our approximate probability of a type I error is α. The exercises below present other examples of contingency tables, as well as other types of tabular summarizations that can be examined in a similar way.

Let us close this section with one further example of a χ^2 test, which is quite different than the ones discussed so far. Many times in applied science, several different research workers investigate the same phenomenon, possibly in quite different ways. Each may, for example, test the same hypothesis, possibly using different continuous test statistics; each one separately may not have enough evidence to reject the hypothesis, but some combined measure may yield ample evidence to reject the hypothesis. Suppose each of k different research workers test the same hypothesis H_0; worker i uses the continuous statistic W_i, $i = 1, 2, \ldots, k$, to make the test. Then, assuming in each case that H_0 is rejected for large values of W_i, define $F_{W_i}(W_i) = U_i$ (if U_i is close to 1, then the ith research worker has observed an extreme result and may want to reject H_0). From the probability integral transform (see Exercise 1.3.15), U_i is a uniform random variable on the interval (0, 1), as is $V_i = 1 - U_i$ ($1 - U_i$ is the significance level of the ith statistic). Assuming that the k workers have independent samples, the V_i's are independent random variables. Then $-2\ ln\ V_i$ is a χ^2 random variable with 2 degrees of freedom (see Exercise 1.3.24) and

$$V = \sum_{i=1}^{k} -2\ ln\ V_i$$

is χ^2 with $2k$ degrees of freedom. If $V \geq \chi_{2k}^2(1 - \alpha)$, the combined weight of all the studies is sufficient to reject H_0, with probability of type I error equal to α. The following example illustrates this test.

Example 5.4.5

Assume that $k = 3$ different medical researchers have tested the hypothesis that American males with high cholesterol diets have the same heart attack rate as those without high cholesterol diets. Assume also that .90, .85, and .88 are the observed percentiles of the continuous statistics used (observed values of the U_i). Note then that none of the three could reject H_0 with $\alpha =$ .05. From a table of natural logarithms we find $\ln .10 = -2.303$, $\ln .15 = -1.897$ and $\ln .12 = -2.120$. Thus the observed value of V is

$$\begin{aligned} v &= -2(-2.303 - 1.897 - 2.120) \\ &= 12.64; \end{aligned}$$

since $\chi_6^2(.95) = 12.592 < 12.64$, we would reject H_0 with $\alpha = .05$, based on the combined evidence of all three investigators.

EXERCISE 5.4

1. Analogous to the proof of Theorem 5.4.1, assume X_1, X_2, X_3, X_4 is a multinomial vector with parameters n, p_1, p_2, p_3, p_4, and show that

$$\sum_{i=1}^{4} \frac{(X_i - np_i)^2}{np_i}$$

is approximately χ_3^2, for large n. Freely make the appropriate approximate normality assumption.

2. A female fly has 50 descendents; if she mated with a male of genotype 1, the expected proportion of flies in each of three classes is 1/3. Given that a particular female produced 20, 20, 10 offspring in the three classes, would you accept the hypothesis she had mated with a male of genotype 1? (Use $\alpha = .05$.)

3. Use the data given in Example 5.4.2 to test the hypothesis that the number of cars abandoned and disposed of, per week, is a Poisson random variable with $\lambda = 1$, versus the alternative that it is Poisson with $\lambda < 1$, using the likelihood ratio test criterion; assume the total number of autos abandoned was 94 in the two years, and use $\alpha = .1$.

4. If **X** is the multinomial vector with parameters $n, p_1, p_2, \ldots, p_k$, show that the *modified minimum* χ^2 estimators of $p_1, p_2, \ldots, p_k$ are identical with the maximum likelihood estimators of these parameters.

5. A random sample of 200 university juniors yielded the following data:

		Sex	
		Male	Female
Grade point	3 or more	40	22
(4 = A)	less than 3	69	69

Test the hypothesis, with $\alpha = .05$, that male and female junior students at this university have equal proportions of students with **B** averages or better.

6. The (generalized) likelihood ratio test criterion can also be used to test independence in a contingency table. Suppose each observation we take falls into exactly one cell of an $r \times c$ table, and, for a random sample of n observations, define X_{ij} to be the number of observations in cell ij, $i = 1, 2, \ldots, r$, $j = 1, 2, \ldots, c$. The probability a given observation falls in cell ij is p_{ij}. If the two methods of classification are independent, $p_{ij} = r_i s_j$. Derive the likelihood ratio test criterion of H_0: $p_{ij} = r_i s_j$, versus the alternative that H_0 is not true.

7. Generalize the χ^2 test of independence to a three-dimensional contingency table with rck cells. How many different independence hypotheses are there and what is the appropriate χ^2 statistic for each?

8. Derive the likelihood ratio test criterion for each of the tests defined in question 7.

9. Tables of random numbers present digits, or sets of digits, in random order. These sequences are subjected to various goodness-of-fit tests to see if they are truly random. Select a page at random in your telephone book and, starting at the left-hand edge of the page, record the final digits of the first 100 telephone numbers. Summarize the number of 0's, 1's, . . . , 9's that you have recorded and test the hypothesis that each digit occurs with probability 1/10.

10. Suppose k different research workers all test the same hypothesis, using continuous statistics $W_1, W_2, \ldots, W_k$. If each will reject H_0 for small values of W_i. what is the appropriate method for combining their evidence?

11. Suppose each of the research workers in question 10 is using a two-tailed test. Let m_i be the median of W_i (thus $P(W_i \le m_i) = .5$) and define

$$\begin{aligned} Y_i &= 2F_{W_i}(W_i) && \text{if } W_i \le m_i \\ &= 2[1 - F_{W_i}(W_i)] && \text{if } W_i > m_i \end{aligned}$$

Show that Y_i is uniform on (0, 1); how would you use $Y_1, Y_2, \ldots, Y_k$ to combine the evidence over the k experiments?

5.5 The Sequential Probability Ratio Test

We shall in this section examine a more modern technique for testing a simple hypothesis versus a simple alternative. As we saw in Sections 5.1 and 5.2, it is always possible to find a test that will make the probability of one of the two types of error equal to zero; thus the usual procedure is to fix the probability of a type I error at α and then to search for that test which has the smallest possible value for the probability of the other type of error (β), for the given fixed sample size n. The Neyman-Pearson fundamental lemma, Theorem 5.2.1, shows that the best critical region for testing a simple hypothesis $\theta = \theta_0$ versus a simple alternative $\theta = \theta_1$ is constructed by putting in the acceptance region A those sample points for which

$$\frac{L(\theta_1)}{L(\theta_0)} < k,$$

where k is chosen to make the probability of type I error equal to α. β then is as small as possible, for n fixed.

During World War II several statisticians suggested that the sample values are in fact observed sequentially, rather than all at once, in many practical problems. Thus, between observations of sample values, time may be available in which to do calculations; in such a case it would seem logical to consider a sequential procedure for testing a simple hypothesis versus a simple alternative. After each observation is made, the experimenter could decide to take one of three possible actions:

a. Accept H_0 (reject H_1),
b. Reject H_0 (accept H_1),
c. Take another sample.

In such a case the sample size required is itself a random variable. The late Abraham Wald described a method for making such a sequential test which is very simple to apply, for given values of α and β, and the expected value of the sample size required is quite small, compared to the fixed size n which would be required to give the most powerful test with the same α and β.

Wald proposed using the *sequential probability ratio test* which proceeds as follows: We have a (population) random variable X with density f_X (or probability function p_X), which depends on the value of a single unknown parameter θ. Since the sample size for the procedure will be a random variable, N, whose possible values will be represented by n, we will find it convenient to denote the likelihood function of the sample, given $N = n$, by $L(\theta, n)$, in discussing the sequential probability ratio test. Wald's procedure is as follows, to test H_0: $\theta = \theta_0$ versus H_1: $\theta = \theta_1$. As observations are made, it is possible to compute $L(\theta_i, 1)$, $L(\theta_i, 2)$, $L(\theta_i, 3), \ldots,$ for $i = 0, 1$ and thus we can also compute the sequence of ratios $[L(\theta_1, 1)]/[L(\theta_0, 1)]$, $[L(\theta_1, 2)]/[L(\theta_0, 2)]$, $[L(\theta_1, 3)]/[L(\theta_0, 3)], \ldots;$ for any pair of constants $k_0 < k_1$ the ratio $[L(\theta_1, n)]/[L(\theta_0, n)]$ will either lie between k_0 and k_1 or it will be smaller than k_0 or bigger than k_1, for every n. The sequential probability ratio test says

a. Accept H_0 if

$$\frac{L(\theta_1, n)}{L(\theta_0, n)} \leq k_0,$$

b. Reject H_0 if

$$\frac{L(\theta_1, n)}{L(\theta_0, n)} \geq k_1,$$

c. Continue sampling if

$$k_0 < \frac{L(\theta_1, n)}{L(\theta_0, n)} < k_1.$$

k_0 and k_1 are chosen to make the probabilities of type I and type II errors equal to α and β, respectively. This procedure can be depicted graphically as given in Figure 5.5.1. As pictured there, sampling would end at $n = 7$, and H_0 would be accepted.

By taking the log of the likelihood ratio, the decision to accept H_0, reject H_0, or continue sampling can generally be made in terms of a statistic whose value can be easily computed (and recomputed) as each sample value becomes available. That is,

$$k_0 < \frac{L(\theta_1, n)}{L(\theta_0, n)} < k_1$$

is equivalent to

$$\log k_0 < \log L(\theta_1, n) - \log L(\theta_0, n) < \log k_1;$$

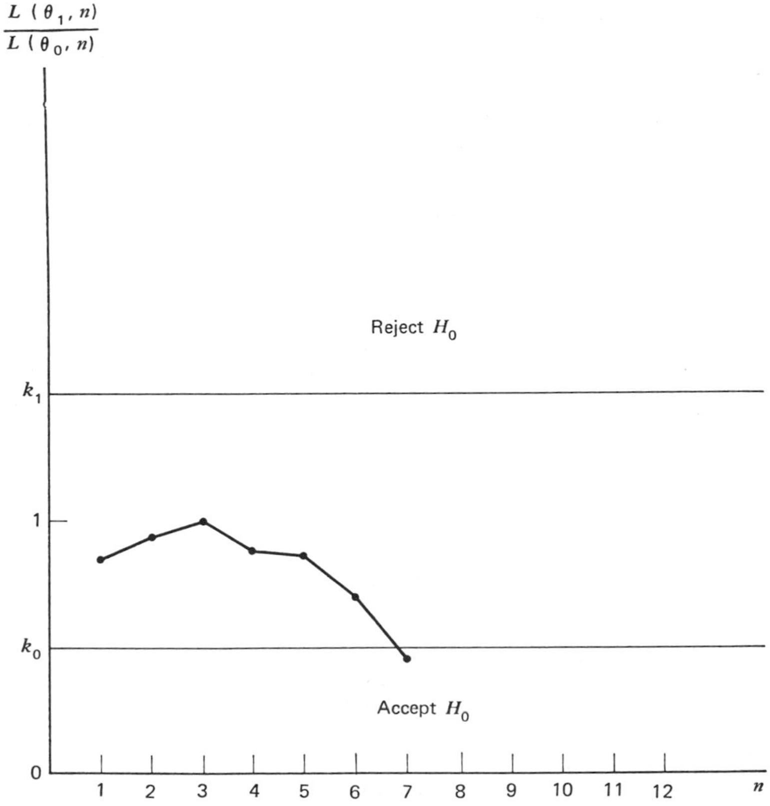

FIGURE 5.5.1. Graph for the sequential probability ratio test.

this inequality can generally be easily manipulated to give the equivalent inequality

$$c_0(n) < g(x_1, x_2, \ldots, x_n) < c_1(n)$$

where $c_0(n)$ and $c_1(n)$ are constants that depend on the sample size n and g is the statistic whose value determines which decision to take. The following example illustrates this equivalence.

Example 5.5.1

People that are in charge of the quality control of incoming material want to screen those lots of material which contain unacceptably large numbers or proportions of defective items. Specifically, suppose a big lot of material is received; each item is either nondefective or defective. If we let p be the proportion of items in the lot that are defective, then we would like to make inferences about the value of p. If we select one item at random from the lot and let $X = 1$ if it is defective and $X = 0$ if it is not, then X is a Bernoulli random variable with parameter p. Thus

$$\begin{aligned} p_X(x) &= 1 - p, \qquad & x = 0 \\ &= p, & x = 1. \end{aligned}$$

So long as the number of items in the lot is very large, the same probability function is appropriate for X_i, $i = 1, 2, \ldots, n$, where the value of X_i is determined by the quality of the ith item selected. Suppose, then, that we want to test H_0: $p = .01$ versus H_1: $p = .10$ (and that our acceptance of the lot will depend upon which of these two hypotheses we accept). The likelihood function, for any given n, then is

$$L(p, n) = \prod_{i=1}^{n} p_X(x_i) = p^{\sum x_i}(1 - p)^{n - \sum x_i}$$

and the sequential probability ratio test says we should continue sampling as long as

$$k_0 < \frac{L(p_1, n)}{L(p_0, n)} < k_1;$$

that is, as long as

$$k_0 < \frac{(.1)^{\sum x_i}(.9)^{n-\sum x_i}}{(.01)^{\sum x_i}(.99)^{n-\sum x_i}} < k_1$$

which is equivalent to

$$k_0 < 10^n \left(\frac{1}{11}\right)^{n-\sum x_i} < k_1$$

or

$$\frac{\log k_0 - n + n \log 11}{\log 11} < \sum x_i < \frac{\log k_1 - n + n \log 11}{\log 11}$$

where the logarithms are taken with base 10. The statistic whose value is used to decide whether to accept H_0 or reject H_0 or to continue sampling then is $g(x_1, x_2, \ldots, x_n) = \sum x_i$, the total number of defectives, and the boundaries between these decisions are given by $c_0(n) = a_0 + bn$, $c_1(n) = a_1 + bn$, where

$$a_0 = \frac{\log k_0}{\log 11}, \qquad a_1 = \frac{\log k_1}{\log 11}, \qquad b = 1 - \frac{1}{\log 11},$$

two parallel straight lines in the $(n, \sum x_i)$ plane. Figure 5.5.2 gives a graph of this procedure. As long as the plot of number of defectives versus sample size stays between the two lines we continue sampling; if it crosses the bottom line $(c_0(n))$, we stop sampling and accept H_0, whereas if it crosses the top line $(c_1(n))$, we stop sampling and reject H_0.

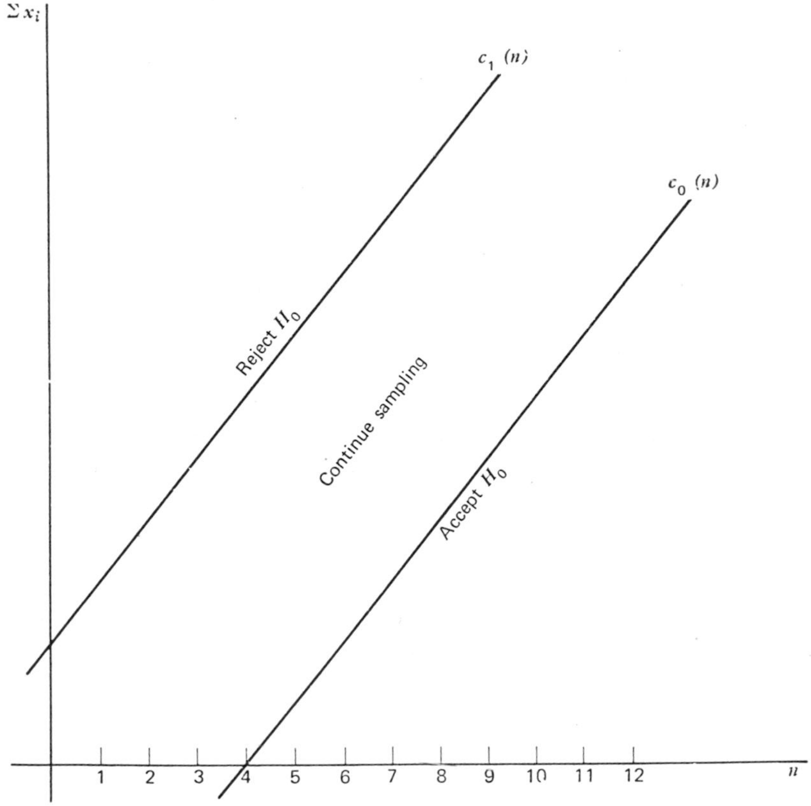

FIGURE 5.5.2. Sequential probability ratio test in terms of $\sum x_i$ (see Example 5.5.1).

By looking at either Figure 5.5.1 or 5.5.2 we can see that there is an unending channel where we should continue sampling; if the likelihood ratio (or the statistic g) remains in this channel, we would not reach a decision about accepting or rejecting H_0. It can be shown that the probability the ratio remains in this channel is 0, as $n \to \infty$. Thus this sequential test will end with probability 1, in either the decision to accept H_0 or to reject H_0. Thus, assuming H_0 is true, in the limit we have probability $1 - \alpha$ of accepting H_0 and probability α of rejecting H_0. Assuming H_1 is true these limiting probabilities are β and $1 - \beta$, respectively.

We have not yet addressed the question of determining the values of k_0 and k_1 so that we have known values for α and β, the probabilities of the two types of error. The determination of the exact values for k_0 and k_1 to give specific values for α and β can be a difficult chore; luckily a good approximation to their values can be determined as follows. We shall present the argument only for the case in which we are sampling from a discrete population but the same approximations are valid in the continuous case as well.

Define B_n to be the set of all sample outcomes $\mathbf{x}$ such that

(*a*) $k_0 < [L(\theta_1, j)]/[L(\theta_0, j)] < k_1, \; j = 1, 2, \ldots, n-1$

(*b*) $[L(\theta_1, n)]/[L(\theta_0, n)] \geq k_1$.

Thus, B_n is the collection of all sample outcomes such that we would reject H_0 (accept H_1) on the nth sample value and not before. For each such sample, then, $L(\theta_1, n) \geq k_1 L(\theta_0, n)$.

In sampling from a discrete population we recall that $L(\theta, n)$ evaluated at the sample values actually gives the probability of the sample values occurring. The probability assigned to $\bigcup_{n=1}^{\infty} B_n$ (the totality of all samples $\mathbf{x}$ leading to rejection of H_0) is

$$\sum_{n=1}^{\infty} \sum_{\mathbf{x} \in B_n} L(\theta, n)$$

for any specific value of θ. Note then that we have

$$\sum_{n=1}^{\infty} \sum_{\mathbf{x} \in B_n} L(\theta_1, n) \geq k_1 \sum_{n=1}^{\infty} \sum_{\mathbf{x} \in B_n} L(\theta_0, n)$$

or

$$1 - \beta \geq k_1 \alpha$$

since the probability we accept H_1, given it is true is $1 - \beta$, and the probability we reject H_0 given it is true is α. Thus

$$k_1 \leq \frac{1 - \beta}{\alpha}$$

and we have a very simple, but very useful, upper bound on k_1.

Similarly, let A_n be the set of all possible sample outcomes such that we accept H_0 on observing the nth sample value. That is, A_n is the set of all sample outcomes $\mathbf{x}$ such that

$$(a)\ k_0 < \frac{L(\theta_1, j)}{L(\theta_0, j)} < k_1, \qquad j = 1, 2, \ldots\ n - 1$$

$$(b)\ \frac{L(\theta_1, n)}{L(\theta_0, n)} \leq k_0.$$

Each such sample $\mathbf{x}$ is at least $1/k_0$ times as likely to occur if H_0 is true than if H_1 is true, since $(1/k_0)L(\theta_1, n) \leq L(\theta_0, n)$, and we have

$$\frac{1}{k_0} \sum_{n=1}^{\infty} \sum_{x \in A_n} L(\theta_1, n) \leq \sum_{n=1}^{\infty} \sum_{x \in A_n} L(\theta_0, n)$$

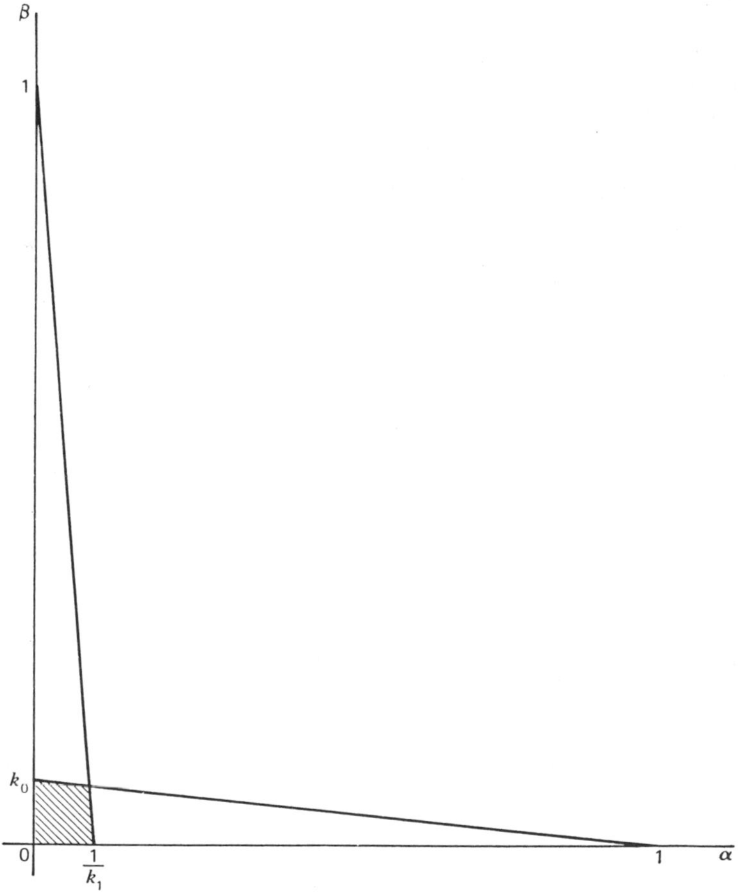

FIGURE 5.5.3. Relationship of α and β to k_0 and k_1.

or

$$\frac{1}{k_0}\beta \leq 1 - \alpha;$$

this gives $k_0 \geq \beta/(1 - \alpha)$ as a lower bound for k_0.

If, then, we let α^*, β^* be two numbers, each between 0 and 1, noninclusive, we can define $k_0 = \beta^*/(1 - \alpha^*)$, $k_1 = (1 - \beta^*)/\alpha^*$. If we use these values for k_0 and k_1 in our sequential probability ratio test we then have $\beta^*/(1 - \alpha^*) \geq \beta/(1 - \alpha)$ and $(1 - \beta^*)/\alpha^* \leq (1 - \beta)/\alpha$. It is easy to see that these two inequalities imply that $\alpha + \beta \leq \alpha^* + \beta^*$; i.e., the sum of the actual probabilities of the two types of error is no larger than the sum of α^* and β^*, used to define k_0 and k_1. Furthermore, $\beta/(1 - \alpha) \leq k_0$ and $\alpha/(1 - \beta) \leq 1/k_1$ imply $\beta \leq k_0(1 - \alpha)$, $\alpha \leq (1 - \beta)/k_1$. Plotting the lines $\beta = k_0(1 - \alpha)$, $k_1\alpha = 1 - \beta$ in the (α, β) plane we can see that the true values of α and β must lie in the quadrilateral that is shaded in Figure 5.5.3.

Example 5.5.2

If in the quality control problem stated in example 5.5.1, we wanted to have approximately $\alpha = .01$ and $\beta = .05$, we should use

$$k_0 = \frac{.05}{.99} = .051, \qquad k_1 = \frac{.95}{.01} = 95.$$

This leads to

$$c_0(n) = \frac{(\log(.051) - n + n \log 11)}{\log 11} = -1.24 + .04n$$

$$c_1(n) = \frac{(\log 95 - n + n \log 11)}{\log 11} = 1.90 + .04n;$$

as long as

$$-1.24 + .04n < \sum x_i < 1.90 + .04n$$

we continue sampling. The first time we find $\sum x_i \leq -1.24 + 0.4n$ we accept H_0 and the first time we find $\sum x_i \geq 1.90 + .04n$ we reject H_0. Our probabilities of error are (approximately) $\alpha = .01$, $\beta = .05$.

As has been mentioned, the sample size needed to reach a decision with the sequential probability ratio test is a random variable. It can be shown that the expected number of observations needed to make a decision about H_0 is approximately given as follows: Let $f_X(x \mid \theta_0)$ denote the density (or probability function) of the random variable sampled, assuming H_0 is true, and let $f_X(x \mid \theta_1)$ denote this same quantity under the assumption that H_1 is true, and define $Z = \log [f_X(X \mid \theta_1)/f_X(X \mid \theta_0)]$. Then the expected number of observations required in using the sequential probability ratio test of H_0: $\theta = \theta_0$ versus H_1: $\theta = \theta_1$ is approximately given by

$$E(N \mid H_0) = \frac{(1 - \alpha)\log k_0 + \alpha \log k_1}{E(Z \mid H_0)}$$

$$E(N \mid H_1) = \frac{\beta \log k_0 + (1 - \beta)\log k_1}{E(Z \mid H_1)}.$$

The following example illustrates the computation of these quantities for the quality control example already introduced.

Example 5.5.3

Referring to the same quality control problem mentioned earlier, if we use the sequential probability ratio test of H_0: $p = .01$ versus H_1: $p = .10$ with $\alpha = .01$ and $\beta = .05$, we have $k_0 = .051$, $k_1 = 95$,

$$Z = \log [(.10)^X(.90)^{1-X}/(.01)^X(.99)^{1-X}] = X \log 11 + \log (10/11).$$

Then

$$E(Z \mid H_0) = (.01) \log 11 + \log (10/11) = -.0310,$$

$$E(Z \mid H_1) = (.10) \log 11 + \log (10/11) = .0627$$

and we have

$$E(N \mid H_0) = \frac{(.99)(-1.297) + (.01)(1.978)}{-.0310}$$

$$= 40.8 \doteq 41$$

$$E(N \mid H_1) = \frac{(.1)(-1.297) + (.9)(1.978)}{.0627}$$

$$= 26.3 \doteq 27.$$

From the Neyman-Pearson lemma, the most powerful fixed sample size critical region for testing H_0: $p = .01$ versus H_1: $p = .10$, with $\alpha = .01$, is of the form $\sum X_i \leq c$; $\sum X_i$ is binomial with parameters n and p and c is chosen to make $\alpha = .01$. We then can choose n such that $\beta = .05$. We find that with a fixed sample size of $n = 78$ we should reject H_0 if the total number of defectives is 4 or more; using this rejection region we have $\alpha \doteq .01$, $\beta \doteq .05$. Note that the sequential probability ratio test has an expected sample size considerably smaller than 78 to achieve (approximately) the same values for α and β.

We have discussed the sequential probability ratio test only in terms of testing a simple hypothesis versus a simple alternative. If we more realistically want to test a one-sided hypothesis H_0: $\theta = \theta_0$ versus the alternative H_1: $\theta > \theta_0$, we must choose some particular value $\theta_1 > \theta_0$, which will be used in sequentially computing the likelihood ratio. The values of k_0 and k_1 are determined as above and the resulting α and β then refer to

$$P(\text{reject } H_0 \mid \theta = \theta_0)$$

and $P(\text{accept } H_0 \mid \theta = \theta_1)$, respectively.

EXERCISE 5.5

1. The example treated in this section is frequently referred to as *attribute sampling*, a case in which each item is classified as being either defective or nondefective. When concerned with actual continuous measurements, the sampling is called *variable sampling*. The following is a problem in variable sampling. Assume that a manufactured part has one critical dimension and that the values for this dimension are normally distributed with unknown mean μ and variance 1. Derive the sequential probability ratio test for H_0: $\mu = 10$ versus H_1; $\mu = 10\ 1/2$, both in terms of k_0 and k_1 and in terms of $c_0(n)$ and $c_1(n)$, with $\alpha = \beta = .05$.

2. Compute the expected sample size needed for the test in 1, under H_0 and H_1, and compare these values with the fixed sample size, most powerful test with $\alpha = \beta = .05$.

3. Assume that the critical dimension of a part made by a given process is normally distributed with mean $\mu = 10$ and unknown σ^2. Derive the sequential probability ratio test of H_0: $\sigma^2 = 1$ versus H_1: $\sigma^2 = 2$, both in terms of k_0 and k_1 and in terms of $c_0(n)$ and $c_1(n)$, with $\alpha = .05$, $\beta = .10$.

4. Compute the expected sample sizes needed for the test in problem 3, under H_0 and H_1, and compare these values with the fixed sample size, most powerful test with $\alpha = .05$, $\beta = .10$.

5. In practical applications both the mean and variance of the normal distribution sampled are generally unknown. What effect does this have on the tests derived in questions 1 and 3? What information would be needed (or assumption that should be made) to carry out the tests?

6. Assume that the number of traffic deaths per week on a 10-mile stretch of highway is a Poisson random variable with parameter λ. Derive the sequential probability ratio test of H_0: $\lambda = 1$ versus H_1: $\lambda = 2$, with $\alpha = \beta = .10$.

7. Compute the expected sample sizes necessary for the test in question 6 and compare with the most powerful fixed sample size test for the same α and β.

8. Suppose that in question 6 we do not have the actual number of deaths per week, but only whether there were no deaths or not, per week. That is, in a given week we have the information that either there were no deaths or there was one or more deaths. Derive the sequential probability ratio test of H_0: $\lambda = 1$ versus H_1: $\lambda = 2$ with $\alpha = \beta = .10$.

9. Compare the expected sample sizes needed in problem 8 with those derived in problem 7.

CHAPTER 6

Decision Theory and Bayesian Methods

6.1 Introduction

Decision theory provides a general structure for expressing and examining many types of problems, including those of estimating unknown parameters and testing hypotheses. It provides a unified view of both these problems and presents other ways of judging how good a given procedure or method may be. In this chapter, we shall look at some of the basic elements of decision theory and see their applications to estimation and tests of hypotheses.

Two basic elements of the decision theory approach are the set Ω of possible states of nature and the set $\mathscr{A}$ of possible actions, any one of which may be taken. The true state of nature is of course unknown; were it known, there would be no problem requiring a solution and the appropriate, optimal course of action would be known. For our purposes, Ω will generally be the set of possible distributions for the population from which we will select a sample. The elements of the action space will depend upon the type of problem we are faced with; if we want to estimate one or more unknown parameters, $\mathscr{A}$ will consist of the set of possible estimates (numbers) we might use and generally in such problems $\mathscr{A} = \Omega$. If we want to test a hypothesis, as in Chapter 5, $\mathscr{A}$ will consist of only two elements, accept or reject the stated hypothesis.

We also assume a known *loss function*, $L(\theta, a)$, which describes the loss we will suffer if in fact the true state of nature is $\theta \in \Omega$ and we take action $a \in \mathscr{A}$. $L(\theta, a) \geq 0$ for all $\theta \in \Omega$, $a \in \mathscr{A}$. The loss function $L(\theta, a)$, for certain $\theta \in \Omega$, could be positive for all actions we might take; thus, if $L(\theta_0, a) > 0$ for all $a \in \mathscr{A}$, no matter which action we take, when θ_0 is the true state, we suffer a positive loss. The *regret function*, $R(\theta, a)$, is for each θ defined by

$$R(\theta, a) = L(\theta, a) - \inf_a L(\theta, a);$$

note then that for each possible θ there is some a^* such that $R(\theta, a^*) = 0$; if we take an action a other than a^* such that $R(\theta, a) > 0$, then we "regret" the unnecessary loss $R(\theta, a) - R(\theta, a^*)$. We shall generally use loss functions that are actually regret functions for statistical problems.

Example 6.1.1

Assume $\Omega = \{\theta_1, \theta_2\}$, $a = \{a_1, a_2, a_3\}$ and that the loss function is as specified below.

	a_1	a_2	a_3
θ_1	2	0	$1\frac{1}{2}$
θ_2	0	3	$1\frac{1}{2}$

Since $L(\theta_1, a_2) = L(\theta_2, a_1) = 0$, note that this is actually a regret function. If we knew that θ_1 were the true state of nature, we clearly should take action a_2, whereas if we knew θ_2 were the true state of nature we should choose action a_1. There is no action which is best for both (all) states of nature, as is generally the case. One conservative procedure which is commonly used is the *minimax procedure;* this is based on the concept of looking for the maximum possible loss we might suffer, for each action, and then choosing that action which minimizes this maximum loss: hence the name minimax. For the given loss function, note that $\max_\theta L(\theta, a_1) = 2$, $\max_\theta L(\theta, a_2) = 3$, $\max_\theta L(\theta, a_3) = 1\frac{1}{2}$. Thus the minimax solution is given by a_3; using this strategy we know we will lose $1\frac{1}{2}$ units whereas if we take actions a_1 or a_2 we may lose more than this amount.

The number of possibilities open to us, in choosing an action for $\mathscr{A}$, can be expanded by using a random mechanism, such as flipping a coin or choosing an integer from a random number table, to identify the action we will take. Such an approach is called a *mixed strategy*, whereas not using such a mechanism and always taking the same action is called a *pure strategy.* Note that if we do use a mixed strategy then the loss that we suffer is a random variable. The loss suffered using a pure strategy is a degenerate random variable with probability 1 of equalling the appropriate value on the loss function. To compare mixed and pure strategies, it is usual to compute the expected value of the loss suffered; the strategy with the smallest expected loss is generally to be preferred. Let us examine the example given above allowing the use of mixed strategies.

Example 6.1.2
For the loss function given in Example 6.1.1, the minimax (pure) strategy was a_3. Clearly the expected loss using a_3 is $1\frac{1}{2}$, whether θ_1 or θ_2 is the true state of nature. We might inquire whether a mixed strategy, using elements a_1 and a_2, could give smaller expected losses for both states of nature. Suppose we choose action a_1 with probability p (and a_2 with probability $1 - p$). Then our expected loss, if θ_1 is the true state of nature, is $2p$; if θ_2 is the true state the expected loss is $3(1 - p)$. Thus if we can choose p such that $2p < 1\frac{1}{2}$ and $3(1 - p) < 1\frac{1}{2}$, that is $1/2 < p < 3/4$, we have a mixed strategy whose expected loss is smaller than that of the minimax (pure) strategy for either state of nature. Thus if we take $p = 2/3$, for example our expected loss is 4/3, if θ_1 is the true state, and it is 1, if θ_2 is the true state. Notice that if we take p smaller than 2/3 the expected loss, for θ_1, will decrease, while that for θ_2 will increase. Choosing p so that the two expected losses are equal defines the *minimax mixed strategy;* if p varies from that value, one of the expected losses must increase, and we would not then have the smallest possible maximum expected loss, over the elements $\theta \in \Omega$. Thus, for the minimax mixed strategy we want $2p = 3(1 - p)$, that is $p = 3/5$, and the expected loss is $1\frac{1}{5}$ for either state of nature.

The preceding examples made no use of a random sample from the population; some authors call this a *no-data decision problem.* For cases of interest to us, we should like to be able to use the results of a sample in choosing the action $a \in \mathscr{A}$ that might be best. Assume that we do have a random sample of n observations $\mathbf{X}' = (X_1, X_2, \ldots, X_n)$ from the population. Then it would seem reasonable to consider the different possible mappings from the sample space S of values for $\mathbf{X}$ to the action space $\mathscr{A}$. Any such mapping simply partitions S into pieces $S_1, S_2, S_3, \ldots,$ such that if we observe $\mathbf{X} \in S_i$ then we take action a_i. By looking at all possible ways of making such a partition we should be able to find that one which is best according to some specified criterion. Note that if we do use such an approach, then the actual action $a \in \mathscr{A}$ which we shall take is itself a random variable (actually the random sample is simply a random mechanism for deriving a mixed strategy), since for any specified partition of S, we will only take action a_i if $\mathbf{X} \in S_i$. Thus again the loss which we will incur is itself a random variable, depending on the outcome of the sampling operation. As with the use of mixed strategies discussed above, the expected loss is used to choose the "best" partition of the sample space. These partitions of S defining mappings into $\mathscr{A}$ are called *decision rules.* The expected value of the loss function, averaging over the possible samples we might observe, is called the *risk function.* Thus the risk function is a function only of the decision rule employed and the true state of nature.

A little more concretely, let $d(\mathbf{X})$ represent a decision function; i.e., $d(\mathbf{X})$ is an element of $\mathscr{A}$, given the sample $\mathbf{X}$. Then d simply specifies which particular $a \in \mathscr{A}$ we should take for any particular observed value of $\mathbf{X}$. The random loss we will incur using d then is $L(\theta, d(\mathbf{X}))$ and the risk function is

$$r(\theta, d) = E_{\mathbf{X}}[L(\theta, d(\mathbf{X}))]$$

where the expectation is taken with respect to $\mathbf{X}$ as is noted by the subscript on the expectation operator. We would like to find that rule d which makes $r(\theta, d)$ as small as possible for each θ. As might be expected it is not generally possible to find a decision function d which minimizes $r(\theta, d)$ for all $\theta \in \Omega$, just as we cannot, in all cases, find a uniformly most powerful test or a minimum variance unbiased estimator of an unknown parameter.

Example 6.1.3

As a simple example of some of these ideas, suppose we want to estimate the probability p of getting a head when a Kennedy half-dollar is flipped. Thus the states of nature are $\Omega = \{p\colon\ 0 \le p \le 1\}$ and the action space is $\mathscr{A} = \Omega$ since presumably we would want to be able to guess any one of the possible values which p might equal. Assume we will have available the results of n flips of the coin; thus the sample space for $\mathbf{X}$ is the collection of 2^n n-tuples, each entry of which is 0 or 1. Let us assume a *squared error loss function*

$$L(p, a) = (p - a)^2$$

where we are using p rather than θ as a generic symbol for an element of Ω. We saw in Chapter 3 that $\bar{X} = (1/n) \sum X_i$ is the minimum variance unbiased estimator of p; let us define decision rule d_1 then by $d_1(\mathbf{X}) = \bar{X}$. For comparative purposes let us also define $d_2(\mathbf{X}) = 1/2$, for all $\mathbf{X}$. Thus d_2 really ignores the sample outcome and says we will take action $a = 1/2$ (give 1/2 as our guess for p) no matter what happens when the coin is flipped. Then the two risk functions are

$$\begin{aligned} r(p, d_1) &= E_{\mathbf{X}}[(p - d_1(\mathbf{X}))^2] \\ &= E_{\mathbf{X}}[(p - \bar{X})^2] \\ &= \frac{p(1-p)}{n}; \\ r(p, d_2) &= E_{\mathbf{X}}[(p - d_2(\mathbf{X}))^2] \\ &= E_{\mathbf{X}}[(p - \tfrac{1}{2})^2] \\ &= (p - \tfrac{1}{2})^2. \end{aligned}$$

Note then that the expected loss using d_1 is smaller than that suffered using d_2 unless $\frac{1}{2} - 1/[2\sqrt{n} + 1] \leq p \leq \frac{1}{2} + 1/[2\sqrt{n} + 1]$. Thus, for some states of nature (essentially all states as $n \to \infty$) d_1 is better than d_2 and for others d_2 is the better. If we really wanted to use d_2, of course, there would not be any point in selecting the sample, since we do not use the outcome in any way. The decision theory approach allows us to easily compare the results of such a priori guesses with the results based on an estimator which uses a random sample, given the form of the loss function.

Let us close this section by examining the Bayesian approach to selecting an action in such problems. The Bayesian approach uses subjective probability to measure the relative degree of belief that particular elements $\theta \in \Omega$ may be the true state of nature. In essentially everything we have studied so far, the probabilities used were interpreted in a frequency sense; they were referring to an experiment which could be repeated an indefinite number of times and, if the probability of occurrence of an event $\mathscr{A}$ was .3, we meant that $\mathscr{A}$ would be expected to occur in about 30% of these repetitions. This type of interpretation of probability is called *objective* or *frequentist*; the numbers called probabilities are measuring relative frequencies of occurrence in repetitions of the basic experiment.

In common English usage, however, probability is frequently used in another, more subjective sense. For example, we have all heard such statements as "it probably will rain tomorrow" or "the chances are three out of five the Yankees will win the pennant this year" or "most likely the bank robbery was an inside job." In each of these cases the individual making the statement is using his own experience and knowledge as the basis for the statement and is not referring to some experiment which can or will be repeated an indefinite number of times. These are all examples of uses of *subjective* probabilities, ones that base their validity strictly on the beliefs of the individuals making them. Thus a subjective probability is measuring a person's "degree of belief" in a proposition, which is not necessarily the same as its long-term frequency of occurrence if, indeed, he is referring to something which could be repeated.

The Bayesian approach, then, uses a subjective probability measure on Ω to state the *a priori* probabilities that any particular $\theta \in \Omega$ may be the true state of nature. If Ω is continuous, we shall assume knowledge of a continuous density $f_\Theta(\theta)$ assigning probabilities to intervals of points $\theta \in \Omega$ and, if Ω is discrete, we assume knowledge of a discrete probability function, $p_\Theta(\theta)$, when using the Bayesian approach. Thus it is assumed that the true state of nature is a random variable Θ whose range is Ω. One other basic change in notation then is also necessary. If we assume that a population of values has a normal distribution with mean θ and variance 1, while the mean itself is uniformly distributed from 0 to 1 (according to our subjective belief), then the density function for a sample from the population will be denoted by $f_{\mathbf{X}|\Theta}(\mathbf{x} \mid \theta)$. Thus, the conditional distribution of the population, given any possible value θ for the mean, is normal with variance 1.

When the state of nature is also treated like a random variable then the risk (expected loss) using any given decision rule d is also a random variable. The *Bayesian risk* function, $B(d)$, is the expected value of $r(\Theta, d)$, with respect to the probability law for Θ; that is

$$B(d) = E_{\Theta}[r(\Theta, d)].$$

This, of course, results in a unique number associated with each decision rule, and we would like to choose that rule d which has the smallest possible value for $B(d)$ if we use a Bayesian solution. Let us reexamine Example 6.1.3 from this Bayesian point of view.

Example 6.1.4

Assume that the probability of getting a head with a Kennedy half-dollar is θ; we believe θ is equally likely to lie anywhere on the open interval (0, 1); thus we assume $f_{\Theta}(\theta) = 1$, $0 < \theta < 1$. Again we consider the two decision functions $d_1(\mathbf{X}) = \bar{X}$ and $d_2(\mathbf{X}) = \frac{1}{2}$, assuming we have available the results of n flips of the coin. As we saw in Example 6.1.2, for any fixed θ, the two values of the risk function were

$$r(\theta, d_1) = \frac{\theta(1 - \theta)}{n}$$

$$r(\theta, d_2) = (\theta - \tfrac{1}{2})^2.$$

Then the Bayesian risks are

$$B(d_1) = E_{\Theta}\left[\frac{\Theta(1 - \Theta)}{n}\right] = \frac{1}{n}\int_0^1 \theta(1 - \theta)\, d\theta = \frac{1}{6n};$$

$$B(d_2) = E_{\Theta}[(\Theta - \tfrac{1}{2})^2] = \int_0^1 (\theta - \tfrac{1}{2})^2\, d\theta = \frac{1}{12}.$$

Notice that with this loss function and with this assumed a priori distribution for θ, d_1 is better than d_2 for $n > 2$, and as n gets large, the risk using d_1 shrinks to 0.

It should be noted that such Bayesian procedures interject a subjective element into the action taken. Two different people with the same loss function and same sample could be led to using different decision rules because they assumed different prior distributions. However, if legitimate prior information is available, this approach allows its use in a natural way. In some problems the best action to take is rather insensitive to the particular prior used.

EXERCISE 6.1

1. Given the loss function

	a_1	a_2
θ_1	2	2
θ_2	3	0

determine the minimax (pure) strategy. Derive the corresponding regret function and determine the minimax (mixed) strategy.

2. When there are only two different possible states of nature, a simple graphical technique may be employed to investigate mixed strategies. Suppose we have k different actions possible $a_1, a_2, \ldots, a_k$. For each a_i we have the two different losses we may suffer, depending on whether θ_1 or θ_2 is the true state of nature. Thus, each action $a_1, a_2, \ldots, a_k$, can be represented by a unique point in a two-dimensional plane, whose coordinates are the two values on the loss function. If we thus label these k points $a_1, a_2, \ldots, a_k$, the straight line between two of them say a_1 and a_2, is given by $pa_1 + (1 - p)a_2$, where $0 \leq p \leq 1$ (now a_1 and a_2 represent the two-dimensional vectors). Thus, the expected losses using a mixture of actions a_1 and a_2 correspond to the different points on this line, where p is the probability of choosing action a_1. Plot the points given by the loss function in Example 6.2.1, and draw the straight lines connecting them. Note that the straight line connecting a_1 and a_2 does represent the expected losses of a mixture of these two actions, as a function of p. This procedure is equally valid for $k > 2$ different states of nature but the pictures are not so easily drawn even for $k = 3$.

3. For the loss function

	a_1	a_2	a_3	a_4
θ_1	3	0	1	2
θ_2	0	3	1	1

plot the four points corresponding to the actions and connect the points by the straight lines representing mixtures. What is the minimax pure strategy? The minimax mixed strategy always has equal values of the expected loss for each state of nature; thus the straight line with slope 1 out of the origin will always intersect the minimax mixed strategy. For the figure just drawn, determine the minimax mixed strategy by finding the place where the straight line with slope 1, from the origin, first intersects one of the straight lines connecting two points.

4. Assume as in Example 6.1.3 that we want to estimate the probability p of getting a head when a coin is flipped. The coin is flipped one time, and we define $X = 0$ if a tail occurs, $X = 1$ if a head occurs. Define $d_1(X) = X$ and $d_2(X) = X/2$. Again assume a squared error loss function; i.e., $L(p, a) = (p - a)^2$. For what range of p is $d_1(X)$ preferred to $d_2(X)$, by comparing their risk functions?

5. A normal population has mean μ and variance 1; it is known that either $\mu = 1$ or $\mu = 2$. Thus $\Omega = \{\mu\colon \ \mu = 1, 2\}$ and to estimate μ we would then use $\mathscr{A} = \Omega = \{1, 2\}$. Show that with equal losses for making an error, a possible minimax rule (no data) is to flip a fair coin and if we get a head, take action $a = 1$. If we take a sample of n observations from the population, show that

$$d(\mathbf{X}) = 1 \qquad \text{if } \bar{X} \leq 1\tfrac{1}{2}$$

$$d(\mathbf{X}) = 2 \qquad \text{if } \bar{X} > 1\tfrac{1}{2}$$

is also a minimax rule. Why might we prefer this second rule to the first?

6. Assume that the set of states of nature, as in Example 6.1.4, is $\Omega = \{p\colon \ 0 \leq p \leq 1\}$, the action space is $\mathscr{A} = \Omega$ and the loss function is $L(p, a) = (a - p)^2$. Assume we have prior knowledge that the parameter is uniform on $(\frac{1}{2} - r, \frac{1}{2} + r)$ where $r < \frac{1}{2}$. What is the Bayesian risk of $d(\mathbf{X}) = \bar{X}$, based on n flips of the coin?

6.2 Bayesian Estimation

Bayesian estimates of parameters, as briefly discussed in Section 6.1, are employed in many instances. If we are interested in estimating an unknown parameter then Ω is the set of possible values for the parameter and $\mathscr{A}$, the action space, is equal to Ω. The loss function is of course still arbitrary, but the most commonly used loss function is *squared error*, $L(\theta, a) = (\theta - a)^2$. For our brief introduction to Bayesian estimation, we shall always use a squared error loss function. The interested reader is referred to Ferguson (6) for a more general exposition.

As discussed in Section 6.1, if we are going to employ a Bayesian technique, we also assume a subjective probability distribution over the states of nature, summarizing our a priori knowledge of the value of the unknown parameter. Using this distribution we can then define the *Bayesian risk function:*

$$B(d) = E_\Theta[r(\Theta, d)].$$

The decision function d which minimizes $B(d)$ is called the *Bayesian estimator* of the unknown parameter. The following example illustrates the computation of the Bayesian estimator.

Example 6.2.1

An industrial production line produces items each of which is either defective or nondefective. After the process has started each day, a sample of items is selected to estimate the proportion of defective items being produced. From past history of the line, the proportion of defectives produced is either .05 or .25; of course, if the latter value appears to be the one at which the line is operating on a given day, the production is stopped and the machines are reset. We assume, from past records, the probability that the proportion of defectives is .05 is .95, and thus that the probability is .05 that the proportion of defectives is .25. To illustrate the computation of the Bayesian estimator from the definition we assume a sample of $n = 2$ items are randomly selected from the line, and we want to estimate the probability of a defective item. Then we have $\Omega = \{.05, .25\} = \mathscr{A}$ and the subjective probability measure on Ω is

$$P(\Theta = .05) = .95, \qquad P(\Theta = .25) = .05.$$

The loss function is squared error. There are three elements in the sample space $S = \{0, 1, 2\}$ and

$$p_{X|\Theta}(x \mid \theta) = \binom{2}{x}\theta^x(1 - \theta)^{2-x}.$$

There are a total of $2^3 = 8$ different decision rules that could be used, as displayed in Table 6.2.1 below:

TABLE 6.2.1. Decision Rules

	Number of defectives in sample		
Decision rule	0	1	2
d_1	.05	.05	.05
d_2	.05	.05	.25
d_3	.05	.25	.05
d_4	.05	.25	.25
d_5	.25	.05	.05
d_6	.25	.05	.25
d_7	.25	.25	.05
d_8	.25	.25	.25

The tabular entry gives the action we will take, given that we observe the number of defectives listed at the head of the given column. Note that decision rules d_1 and d_8 are the pure strategies that ignore the sample outcomes.

Table 6.2.2 displays the random risks associated with each of these rules and Table 6.2.3 gives the Bayesian risks, $B(d) = E_\Theta[r(\Theta, d)]$. The Bayesian rule, or estimator, then, is given by d_2: if we observe 0 or 1 defectives we estimate θ to be .05 (and we do not stop production) whereas if we observe two defectives we estimate θ to be .25 (and we stop production to reset the machines).

TABLE 6.2.2. Values of $r(\Theta, d)$

$\Theta =$	.05	.25
d_1	0	.0400
d_2	.0001	.0375
d_3	.0038	.0250
d_4	.0039	.0225
d_5	.0361	.0175
d_6	.0362	.0150
d_7	.0399	.0025
d_8	.0400	0

The determination of the Bayesian estimator in Example 6.2.1 involves straightforward, but very tedious, numerical computation. Especially if the number of elements in S and in $\mathscr{A}$ gets very large, having to list and examine all possible decision rules, as in Example 6.2.1, would be effectively impossible. Luckily this approach is not necessary; all that needs to be done is to determine the *posterior distribution* of Θ, $F_{\Theta|\mathbf{X}}(t \mid \mathbf{x})$, for the given sample values $\mathbf{x}$, using Bayes' theorem. Straightforward examination of this posterior distribution then yields the Bayesian decision for that given sample outcome, which of course is all that is required. Much less effort is required,

TABLE 6.2.3. Values of $B(d)$

d_1	d_2	d_3	d_4
.00200	.00197	.00486	.00483
d_5	d_6	d_7	d_8
.03517	.03514	.03803	.03800

in general, in making this determination than is necessary in the complete listing of all possible decision rules. This result is given in Theorem 6.2.1.

THEOREM 6.2.1. Let $\mathbf{X}' = (X_1, X_2, \ldots, X_n)$ be a random sample from a population whose distribution depends on an unknown parameter θ, let Ω be the set of possible values for θ and let $\mathscr{A} = \Omega$. Then if the rule d^* minimizes $E_{\Theta|\mathbf{X}}[L(\Theta, d) \mid \mathbf{X}]$, for each possible sample vector $\mathbf{x}$, d^* is the Bayesian rule. The expectation is taken with respect to the posterior distribution of Θ.

Proof: We shall examine the proof in the case that $F_\Theta(t)$ and $F_{\mathbf{X}|\Theta}(\mathbf{t} \mid \theta)$ are both continuous. The proof is similar for the other cases. By definition, the Bayesian rule d^* satisfies $B(d^*) \leq B(d)$, for all possible rules d. Also

$$B(d) = \int_\Omega r(\theta, d) f_\Theta(\theta)\, d\theta$$

$$= \int_\Omega \left\{ \int_{R_\mathbf{X}} L(\theta, d) f_{\mathbf{X}|\Theta}(\mathbf{x} \mid \theta)\, d\mathbf{x} \right\} f_\Theta(\theta)\, d\theta.$$

From conditional probability we know that

$$f_{\mathbf{X}|\Theta}(\mathbf{x} \mid \theta) f_\Theta(\theta) = f_{\mathbf{X},\Theta}(\mathbf{x}, \theta) = f_{\Theta|\mathbf{X}}(\theta \mid \mathbf{x}) f_\mathbf{X}(\mathbf{x});$$

thus interchanging the order of integration we can write

$$B(d) = \int_{R_\mathbf{X}} \left\{ \int_\Omega L(\theta, d) f_{\Theta|\mathbf{X}}(\theta \mid \mathbf{x})\, d\theta \right\} f_\mathbf{X}(\mathbf{x})\, d\mathbf{x}.$$

Clearly, if for each $\mathbf{x}$ we find d^* such that

$$\int_\Omega L(\theta, d^*) f_{\Theta|\mathbf{X}}(\theta \mid \mathbf{x})\, d\theta \leq \int_\Omega L(\theta, d) f_{\Theta|\mathbf{X}}(\theta \mid \mathbf{x})\, d\theta,$$

that is $E_{\Theta|\mathbf{X}}[L(\Theta, d^*)] \leq E_{\Theta|\mathbf{X}}[L(\Theta, d)]$, then $B(d^*) \leq B(d)$ and d^* is the Bayesian decision rule. Let us immediately apply this result to Example 6.2.1.

Example 6.2.2

Assume as before that $\Omega = \mathscr{A} = \{.05, .25\}$, $P(\Theta = .05) = .95$, $P(\Theta = .25) = .05$, that the number X of defectives in the sample of $n = 2$, given $\Theta = \theta$, is binomial with parameters 2 and θ, and the loss function is squared error. If in the sample we observe $x = 0$ defectives, the posterior distribution of Θ is

$$P(\Theta = .05 \mid 0) = \frac{.95(.95)^2}{.95(.95)^2 + .05(.75)^2} = .967$$

$$P(\Theta = .95 \mid 0) = \frac{.05(.75)^2}{.95(.95)^2 + .05(.75)^2} = .033.$$

The only possible choices with $x = 0$ is to choose $\theta = .05$ or $.25$, since $\mathscr{A} = \Omega$. Since

$$E[(\Theta - .05)^2 \mid X = 0] = 0(.967) + (.04)(.033) = .00132$$

$$E[(\Theta - .25)^2 \mid X = 0] = (.04)(.967) + 0(.033) = .03868,$$

the Bayesian decision rule, with $x = 0$, decides $\theta = .05$. Similarly, by examining the posterior distribution given $X = 1$ and then $X = 2$, we simply reproduce rule d_2 from Example 6.2.1, which we found there to be the Bayesian estimator.

In many important cases the following theorem can be employed, which gives a very simple way of determining the Bayesian rule. The proof will be left for the reader.

THEOREM 6.2.2. If the loss function is squared error, $L(\theta, d) = (\theta - d)^2$, then the Bayes estimator of θ is given by the mean of the posterior distribution, provided that this mean value is a member of $\mathscr{A}$ for each **X**.

Let us close this section with two examples of the use of Theorem 6.2.2.

Example 6.2.3

In an ecological study of a certain lake, it was desired to estimate the proportion of the fish of a given type in the lake which had some evidence of mercury poisoning; let this unknown proportion be p. Before examining any fish, the investigators were completely ignorant about the value of p. Thus, they were willing to assume that p might lie anywhere between 0 and 1 and that no values in this interval were preferred over others. In short, they were willing to assume subjectively that P (the random variable) was uniform on (0, 1), and thus $f_P(p) = 1$, $0 < p < 1$. They then selected a sample of n fish of this type from the lake; for simplicity we assume they were selected independently and at random. (Why might this not be a realistic assumption?) Then the distribution of X, the total number of fish in the sample exhibiting evidence of mercury poisoning in the sample, given $P = p$, is

$$p(X = x \mid p) = \binom{n}{x} p^x(1 - p)^{n-x}, \qquad x = 0, 1, \ldots, n.$$

Let us use Theorem 6.2.2 to derive the Bayes estimator for p. The marginal distribution of X is

$$\begin{aligned} P(X = x) &= \int_0^1 \binom{n}{x} p^x(1 - p)^{n-x}\, dp \\ &= \binom{n}{x} \frac{x!(n - x)!}{(n + 1)!} = \frac{1}{n + 1}, \qquad x = 0, 1, 2, \ldots, n \end{aligned}$$

and the posterior density for P is

$$f_{P|X}(p \mid x) = \frac{\Gamma(n + 2)}{\Gamma(x + 1)\Gamma(n - x + 1)} p^x(1 - p)^{n-x}, \qquad 0 < p < 1,$$

a beta density with parameters $x + 1$ and $n - x + 1$. From Chapter 1 we know the mean of this beta density is $(x + 1)/(n + 2)$ and thus the Bayesian estimator for p is $(X + 1)/(n + 2)$. Both the method of moments and maximum likelihood yield X/n as the estimator in this case; notice that the Bayesian estimator has essentially increased the size of the sample by 2 and the number of observed successes by 1. It is also of interest to note that the mode (maximum) of this posterior distribution is at x/n and thus the maximum likelihood estimator corresponds to the mode rather than the mean of the posterior, in this case. See Exercise 6.2.3 for the reason this occurs.

In the preceding example we assumed what some authors call a *conjunctive prior*, a prior distribution which comes from the same general family as the sample. Looking at $p^x(1-p)^{n-x}$, ignoring the constant, where p is the variable, we see that the sample distribution is of the beta type; the uniform prior which we assumed is also a particular beta. Use of such a related prior frequently is advantageous for subsequent computations. In the following example we again use a conjunctive prior.

Example 6.2.4
A manufacturer of transistors is asked to produce a transistor to new specifications, for use in a specific environment. From previous experience the manufacturer is sure the time to failure, T, of a transistor of this type will be an exponential random variable with parameter λ and he feels 80% sure that λ will be no larger than .01. He decides to investigate the Bayesian estimator of λ, based on a random sample of n lifetimes, and will use a conjunctive prior to represent his subjective feelings about the value of λ. Thus he assumes as a prior for Λ

$$f_\Lambda(\lambda) = 160e^{-160\lambda}, \qquad \lambda > 0$$

since $\int_0^{.01} 160e^{-160\lambda}\, d\lambda \doteq .80$. The conditional density of the n sample lifetimes, given $\Lambda = \lambda$, is

$$f_{\mathbf{T}|\Lambda}(t \mid \lambda) = \lambda^n e^{-\lambda \sum t_i}.$$

The marginal distribution of $\mathbf{T}$ then is

$$\begin{aligned} f_{\mathbf{T}}(\mathbf{t}) &= \int_0^\infty 160\lambda^n e^{-\lambda(\sum t_i + 160)}\, d\lambda \\ &= 160 \frac{\Gamma(n+1)}{(\sum t_i + 160)^{n+1}} \end{aligned}$$

and the posterior density for Λ is

$$f_{\Lambda|\mathbf{T}}(\lambda \mid \mathbf{t}) = \frac{1}{\Gamma(n+1)} [\textstyle\sum t_i + 160]^{n+1} \lambda^n e^{-\lambda(\sum t_i + 160)}.$$

The mean value of this gamma density is $(n+1)/(\sum t_i + 160)$ and thus the Bayes' estimator for λ is $(n+1)/(\sum T_i + 160)$. The maximum likelihood estimator is $n/\sum T_i$. Thus in this case the Bayes' estimator has increased the sum of the failure times by 160 (the value of the parameter in the exponential prior) and increased the number of items tested by 1; thus maximum likelihood with a sample size 1 larger, and a value of 160 for that additional sample, gives the same estimate as Bayes with a sample of n.

EXERCISE 6.2

1. In Example 6.2.1, find the Bayes rule if the proportion of defectives is equally likely to be .05 or .25.

2. In Example 6.2.2, derive the posterior distribution given $X = 1$ and given $X = 2$, and verify that the Bayes estimator takes on values .05 and .25, respectively, in these two cases.

3. Prove that if we have a uniform prior on (a, b) for an unknown parameter, then the maximum likelihood estimator is equal to the mode of the posterior distribution.

4. Prove Theorem 6.2.2.

5. Prove that if our loss function is the absolute value of the error,

$$L(\theta, d) = |\theta - d|,$$

then the Bayes estimator is the median of the posterior distribution.

6. Show that if we are absolutely certain of the value of a parameter, then the Bayes estimator will equal that certain value no matter what we observe in a sample. (That is, if our prior is $P(\Theta = \theta_0) = 1$, then the Bayes estimator equals θ_0 with probability 1; thus a closed mind cannot be changed no matter what happens on the sample.)

7. Assume that X is uniform on the interval $(0, \theta)$ and that θ has the gamma prior

$$f_\Theta(\theta) = \theta e^{-\theta}, \qquad \theta > 0.$$

For a sample of $n = 1$ observation, find the Bayes estimator for θ.

8. Assume that X is a geometric random variable with parameter p and that the prior for p is uniform on (0, 1). Find the Bayes estimator for p, based on a random sample of n observations of X.

9. Generalize problem 8 so that the prior for p is a beta random variable with known parameters α_0 and β_0 (see Section 1.3). What is the Bayes estimator for p?

10. Assume that X is a normal random variable with known variance b and unknown mean μ. The prior density for μ is normal with known mean μ_0 and known variance σ_0^2. Given a random sample of n observations of X, show that the Bayes estimator for μ is

$$\left(\frac{n\bar{X}}{b} + \frac{\mu_0}{\sigma_0^2}\right) \Big/ \left(\frac{n}{b} + \frac{1}{\sigma_0^2}\right).$$

As our prior gets more diffuse, that is as σ_0^2 increases without bound, what happens to this estimator? As our prior gets sharper, that is as $\sigma_0^2 \to 0$, what happens to this estimator?

6.3 Intervals and Tests

We studied confidence intervals in Chapter 4; it will be recalled that a confidence interval for an unknown parameter is a pair of statistics L_1 and L_2 such that the probability is at least $(1 - \alpha)$ that the interval (L_1, L_2) includes the true unknown value of θ. This probability statement is based on the distribution of the two random variables L_1 and L_2 for the given sample size and θ is simply an unknown constant. Confidence intervals are objective interval estimates since the probability used is frequentist in nature; over indefinitely repeated samples of the same size from the same population, the proportion of the time that (L_1, L_2) includes θ is $1 - \alpha$.

It is easy to see that very analogous manipulations can be made with a Bayesian approach, using subjective probabilities. Before any sampling is done, the Bayesian approach assumes an a priori, subjective distribution on the parameter space. This distribution could, of course, be used to derive an interval statement about the unknown parameter, without taking a sample of values from the population. If we are, in fact, going to have a set of sample values available, though, it is much more reasonable to use the posterior distribution of the parameter to construct such an interval, since the posterior distribution summarizes both the prior information and the information contained in the sample about the parameter. If the posterior density is symmetric it would generally seem reasonable to center a two-sided interval at the posterior mean or to find the limits b_1, b_2 such that

$$P(\Theta \leq b_1 \mid \mathbf{X}) = P(\Theta \geq b_2 \mid \mathbf{X}) = \frac{\alpha}{2}.$$

The following example illustrates the computation of such a Bayesian interval.

Example 6.3.1

In Example 6.2.3, the ecological study of a lake, the prior distribution of the proportion p of fish exhibiting mercury poisoning was assumed to be uniform on the interval (0, 1) and the posterior distribution was

$$f_{P|X}(p \mid x) = \frac{\Gamma(n + 2)}{\Gamma(x + 1)\Gamma(n - x + 1)} p^x(1 - p)^{n-x}, \qquad 0 < p < 1,$$

a beta density. If, for example, we wanted to use this posterior density to compute a $100(1 - \alpha)\%$ Bayes interval for p, one such solution is given by finding b_1, and b_2 such that

$$\int_0^{b_1} f_{P|X}(p \mid x)\,dp = \int_{b_2}^{1} f_{P|X}(p \mid x)\,dp = \frac{\alpha}{2};$$

recalling the result given in Exercise 1.3.14, linking the beta distribution function and the binomial, note that these requirements for b_1 and b_2 are equivalent to

$$\sum_{i=x+1}^{n+1} \binom{n+1}{i} b_1^i (1-b_1)^{n+1-i} = \frac{\alpha}{2},$$

$$\sum_{i=0}^{x} \binom{n+1}{i} b_2^i (1-b_2)^{n+1-i} = \frac{\alpha}{2}.$$

The values of b_1 and b_2 so determined have a (subjective) posterior probability of $1-\alpha$ of including p between them. The reader should note the great similarity between these equations and those derived in Example 4.3.2 for a $100(1-\alpha)\%$ confidence interval for p. The interpretation of the two is considerably different, however; the confidence interval for p will include the true value of p with (at least) probability $1-\alpha$ over repeated samples, whereas the Bayesian interval for p merely includes $100(1-\alpha)\%$ of the posterior (subjective) distribution for p. The probability that the Bayesian interval covers p in repeated samples may or may not be at least $1-\alpha$, for all possible p.

As with Bayesian estimates of parameters, the choice of the prior distribution can have a considerable impact on the resulting Bayesian interval. Generally, if the prior variance is fairly large, indicating fairly imprecise prior information, and if the sample is fairly large, the resulting Bayesian interval will be quite close to the corresponding confidence interval for the parameter (as in the preceding example). If, however, the prior variance is quite small, the two techniques can lead to considerably different intervals, in length as well as location; in the extreme case of zero prior variance, we are 100% sure of the value of the parameter in using the posterior distribution with a 0 length interval (as we saw in Exercise 6.2.6).

Tests of simple hypotheses can also be easily examined in the decision theory framework and Bayesian methods applied. If we want to test the simple hypothesis that an unknown parameter has value θ_0 versus the simple alternative that it has value θ_1, the parameter space Ω and the action space $\mathscr{A}$ have only two elements: $\Omega = \mathscr{A} = \{\theta_0, \theta_1\}$. The loss function for a correct decision would reasonably be taken to be zero; thus $L(\theta_0, \theta_0) = L(\theta_1, \theta_1) = 0$. To simplify our notation a little, let $L_{10} > 0$ be the loss suffered if we accept $H_0\colon\ \theta = \theta_0$ when in fact $H_1\colon\ \theta = \theta_1$ is true (type II error) and let $L_{01} > 0$ be the loss suffered if we accept $H_1\colon\ \theta = \theta_1$ when

H_0: $\theta = \theta_0$ is true (type I error). Again, to find the Bayesian procedure, we assume we have a prior probability distribution on Ω; let p_0 be our prior probability that $\Theta = \theta_0$ and $p_1 = 1 - p_0$ be the prior probability that $\Theta = \theta_1$. The following theorem shows that the Bayesian procedure (critical region) is of exactly the same form as the most powerful test (given in the Neyman-Pearson lemma), except that the value of k is uniquely specified.

THEOREM 6.3.1. Assume that $\Omega = \mathscr{A} = \{\theta_0, \theta_1\}$, $P(\Theta = \theta_0) = p_0$, $P(\Theta = \theta_1) = p_1$, $L(\theta_0, \theta_0) = L(\theta_1, \theta_1) = 0$, $L(\theta_1, \theta_0) = L_{10}$, $L(\theta_0, \theta_1) = L_{01}$. Then, given a random sample of n values of X, the Bayesian decision rule is of the form: Decide $\theta = \theta_0$ if and only if

$$\frac{L(\theta_1)}{L(\theta_0)} < k = \frac{p_0 L_{01}}{p_1 L_{10}}$$

where $L(\theta)$ is the likelihood function of the sample values.

Proof: Using $L(\theta)$ to represent the sample likelihood function, as in Section 3.2, the marginal distribution of $\mathbf{X}$, the sample vector, is $p_0 L(\theta_0) + p_1 L(\theta_1)$ and the posterior probability distribution for Θ, given $\mathbf{X}$, is

$$p(\Theta = \theta_i \mid \mathbf{X}) = \frac{p_i L(\theta_i)}{p_0 L(\theta_0) + p_1 L(\theta_1)}, \qquad i = 0, 1.$$

If we decide $\Theta = \theta_0$, using decision rule d, the expected posterior loss is

$$E_{\Theta|\mathbf{X}}[L(\Theta, \theta_0) \mid \mathbf{X}] = \frac{L_{10} p_1 L(\theta_1)}{p_0 L(\theta_0) + p_1 L(\theta_1)}$$

since $L_{00} = 0$, and if we decide $\Theta = \theta_1$, it is

$$E_{\Theta|\mathbf{X}}[L(\Theta, \theta_1) \mid \mathbf{X}] = \frac{L_{01} p_0 L(\theta_0)}{p_0 L(\theta_0) + p_1 L(\theta_1)}.$$

As we saw in Theorem 6.2.1, the Bayes rule d^* is the one which minimizes this posterior expected loss. Thus, if

$$\frac{L_{10} p_1 L(\theta_1)}{p_0 L(\theta_0) + p_1 L(\theta_1)} < \frac{L_{01} p_0 L(\theta_0)}{p_0 L(\theta_0) + p_1 L(\theta_1)},$$

that is, if

$$\frac{L(\theta_1)}{L(\theta_0)} < \frac{p_0}{p_1} \frac{L_{01}}{L_{10}},$$

we should decide $\theta = \theta_0$; if the inequality is not true, we should decide $\theta = \theta_1$. Thus the Bayes rule for testing H_0: $\theta = \theta_0$ versus H_1: $\theta = \theta_1$ is identical in form with the most powerful critical region given by the Neyman-Pearson lemma.

Recall from the Neyman-Pearson lemma (Theorem 5.2.1) that we should accept H_0 if $(L(\theta_1))/(L(\theta_0)) < k$, where k is chosen to make the probability of type I error equal to any desired value α. The Bayes procedure actually specifies k, and thus α, and does not leave that quantity to choice.

EXERCISE 6.3

1. A soft drink manufacturer prepares his beverage in large batches. When a batch is finished it is transferred to a bottling machine which then automatically fills and caps bottles. The bottles are each supposed to be filled with 10 ounces of liquid, but of course there is some variability of content from one bottle to another. Let X be the true content of a bottle selected at random and assume X is normal with mean μ and $\sigma = 1/8$ ounce. Each time a new batch is connected to the bottling machine, controls are set on the machine; due to variations in these settings, as well as variations in the specific gravity of the batches bottled, the true mean content per bottle varies from one batch to the next. Previous experience indicates that this variation in mean level from one batch to the next is well approximated by a normal random variable with mean 10 ounces and standard deviation .05 ounce. Assume a new batch has just been transferred to the bottling machine, a random sample of n bottles are selected, and their contents measured. Use the information on variability of true mean content from one batch to the next to construct a prior distribution for the mean value of X. What is the posterior distribution for this mean value, given the sample? Construct a $100(1 - \alpha)\%$ Bayesian interval for this mean value.

2. A random variable X is uniform on the interval $(-\theta, \theta)$ and Θ is uniform on $(1/2, 1)$ (prior). Given a random sample of 1 observation of X, what is the posterior of Θ? Construct the shortest $100(1 - \alpha)\%$ Bayesian interval for θ, using this posterior.

3. Assume that X, given θ, is uniform $(-\theta, \theta)$ and Θ is a gamma random variable with prior density

$$f_\Theta(\theta) = \theta e^{-\theta}, \qquad \theta > 0.$$

Find the posterior density for Θ, given a sample of size 1 of X, and use it to construct the shortest possible $100(1 - \alpha)\%$ interval for θ.

4. In Exercise 6.3.1 above, assume $n = 10$; what is the Bayesian critical region for testing H_0: $\mu = 10$ versus H_1: $\mu = 9.9$, assuming $L_{01} = 1$, $L_{10} = 2$? What are the probabilities of the two types of error? (Take p_0/p_1 to be the ratio of the prior densities at 10, 9.9, respectively.)

5. Assume that a random sample of n values of X, given $\Theta = \theta$, has density $f_{\mathbf{X}|\Theta}(\mathbf{x} \mid \theta)$, and that Θ has prior continuous density $f_\Theta(\theta)$. What is the form of the critical region for the Bayes' rule in testing H_0: $\theta \le \theta_0$ versus H_1: $\theta > \theta_0$?

6. A female fruit fly may have mated with either of two types of male. In either case, each of her offspring will exhibit one of two types of physical characteristics.

If she mated with a male of type 1, the number of offspring having characteristic 1 is binomial, n, p_1, whereas if she mated with a male of type 2, the number of her offspring having characteristic 1 is binomial n, p_2. Assuming $L_{01} = L_{10}$ and equal a priori probabilities that she mated with either type of male, what is the form of the Bayesian test of the hypothesis that she mated with a male of type I versus the alternative that she mated with a male of type 2?

CHAPTER 7

Linear Models

7.1 Introduction

One of the oldest, and still most useful, types of models employed in many areas of scientific work is what is called a *linear statistical model.* A glance through modern journals of psychology, education, social sciences, life sciences, engineering, chemistry, etc., will reveal a seemingly endless array of practical problems in which linear statistical models have been employed. We shall examine this type of model in this chapter.

In general, a linear statistical model assumes and uses *concomitant variables* (or additional information) to aid in the description of the behavior of the variable of primary interest. Let us consider a simple example of this sort. Medical doctors like to have charts or records available that describe the "normal" weights for infants; by having such a concrete reference of "normal progress," it is easier for them to identify and to attempt to diagnose illnesses or malfunctions of an infant. A linear statistical model would be a likely procedure to use in the construction of a model for normal progress of infant weight.

It would seem logical that the weight of an infant will depend upon, or be affected by, its age as well as its length (among a possibly much larger collection of other factors which we shall ignore for simplicity). In fact, if we restrict the age to between one and six months, say, it might be reasonable to assume that the expected or average weight of a baby should change linearly with these two variables. This is a typical assumption made in using a linear statistical model: The expected value or population mean is dependent upon the values of one or more additional variables which can be measured (the dependence on these variables is linear in the unknown parameters). A sample of "healthy" babies could be selected, and the weight, age, and length determined for each; then the techniques of this chapter could be applied to estimate the unknown parameters of the assumed relationship. We would then have available an equation which, given the age and length of a baby, could be used to predict the expected weight of a healthy baby with this age and length. If the actual weight of the baby being examined differs substantially from this prediction, it may indicate some sort of problem.

As another example, the average salary or expected salary that an individual receives may very well depend upon the number of years of formal education he received. A sample of individuals could be selected and the hypothesis tested that such a relationship appears to exist (and if so, the relationship could be estimated). Such models prove useful in many problems in economics.

For a final example, suppose we were interested in the relationship between a person's speed of accomplishing a certain simple task, such as moving his foot from the accelerator to the brake pedal of an automobile, and certain rates of consumption of an alcoholic beverage. We might use a group of, say, mn volunteers and for each measure their speed of accomplishing the task before consuming the beverage. Then, if we are interested in m different rates of consumption of the beverage, we could randomly assign n volunteers to each of the specified rates, have them consume the beverage at the specified rates, and then again measure their speed in accomplishing the task. By analyzing the data, we could investigate whether there does seem to be any difference in speed, presumably caused by the consumption of the beverage, and additionally whether the different rates of consumption seem to affect the speed differently. If desired, we could also investigate the form of the relationship between the speed and the rate. All of these problems can be profitably studied using the techniques of linear statistical models.

It is interesting that, for estimation problems, very minimal distributional assumptions are needed when *least squares* is applied; in fact, if the additional assumption of a normal distribution is made, and maximum likelihood is applied, exactly the same estimators are derived. Thus the estimators we shall derive here are "good," from two distinct points of view. For testing hypotheses of various sorts, it is of course necessary to make assumptions about the distributions of the random variables observed, to compute probabilities of type I and type II errors. Making the assumption of normality leads to techniques that are very easily applied and interpreted. We shall establish some preliminary results in this section and in succeeding sections go on to examine least squares estimation, (normal) maximum likelihood estimation, confidence intervals and tests of hypotheses for regression, and experimental design models.

Let us first review some results about the multivariate normal that were mentioned or derived in Chapters 1 and 2. First if a $p \times 1$ vector $\mathbf{X}$ has the multivariate normal distribution, the only parameters of its density are the mean vector $\boldsymbol{\mu}$ and the variance covariance matrix $\boldsymbol{\Sigma}$. Its moment generating function is

$$m_{\mathbf{X}}(\mathbf{t}) = e^{\mathbf{t}'\boldsymbol{\mu}+(1/2)\mathbf{t}'\boldsymbol{\Sigma}\mathbf{t}}.$$

Any subset of the components of $\mathbf{X}$ have a (marginal) joint normal distribution whose mean vector consists of the appropriate components from $\boldsymbol{\mu}$ and whose covariance matrix is the appropriate submatrix of $\boldsymbol{\Sigma}$. The components of $\mathbf{X}$ are completely independent if and only if $\boldsymbol{\Sigma}$ is a diagonal matrix. If $\mathbf{Y} = \mathbf{DX}$, where the rows of $\mathbf{D}$ are linearly independent, then $\mathbf{Y}$ is multivariate normal with mean vector $\mathbf{D}\boldsymbol{\mu}$ and covariance matrix $\mathbf{D}\boldsymbol{\Sigma}\mathbf{D}'$. If $\mathbf{X}$ is multivariate normal with covariance $\boldsymbol{\Sigma}$, we can always find a transformation $\mathbf{Y} = \mathbf{CX}$, such that $\mathbf{Y}$ has covariance matrix $\mathbf{I}$; in fact $\boldsymbol{\Sigma} = (\mathbf{C}'\mathbf{C})^{-1}$ in this case. If $\mathbf{X}$ is p-variate normal with parameters $\boldsymbol{\mu}$ and $\boldsymbol{\Sigma}$, then $(\mathbf{X} - \boldsymbol{\mu})'\boldsymbol{\Sigma}^{-1}(\mathbf{X} - \boldsymbol{\mu})$ is a χ_p^2 random variable.

$(\mathbf{X} - \boldsymbol{\mu})'\boldsymbol{\Sigma}^{-1}(\mathbf{X} - \boldsymbol{\mu})$ is called a *quadratic form* in $\mathbf{X} - \boldsymbol{\mu}$, since it consists of a weighted sum of the squares and crossproducts of the components of $\mathbf{X} - \boldsymbol{\mu}$. $\boldsymbol{\Sigma}^{-1}$ is called the matrix of the quadratic form. Since $\boldsymbol{\Sigma}^{-1}$ is positive definite, the quadratic form is positive for all $\mathbf{X} - \boldsymbol{\mu} \neq \mathbf{0}$. We shall need several additional results about quadratic forms for our study of linear models. We shall derive these results now.

Many of the quadratic forms we shall investigate have *idempotent matrices*. A matrix $\mathbf{A}$ is idempotent if and only if $\mathbf{A}^2 = \mathbf{A}$. Since our idempotent matrices occur in quadratic forms, they will also be symmetric. There are a number of properties that symmetric idempotent matrices possess that are extremely useful for our purposes. Those we shall use frequently are:

1. $\mathbf{A}$ is symmetric and idempotent implies every nonzero characteristic root equals 1.
2. The trace of $\mathbf{A}$ equals the rank of $\mathbf{A}$,

$$\operatorname{tr}(\mathbf{A}) = r(\mathbf{A}).$$

3. If $\mathbf{A}$ is idempotent, so is $\mathbf{I} - \mathbf{A}$.

Some simple examples of symmetric idempotent matrices are

$$\begin{pmatrix} \frac{1}{2} & \frac{1}{2} \\ \frac{1}{2} & \frac{1}{2} \end{pmatrix}, \quad \begin{pmatrix} \frac{2}{3} & -\frac{1}{3} & -\frac{1}{3} \\ -\frac{1}{3} & \frac{2}{3} & -\frac{1}{3} \\ -\frac{1}{3} & -\frac{1}{3} & \frac{2}{3} \end{pmatrix}, \quad \frac{1}{n}\mathbf{J},$$

where $\mathbf{J} = \|1\|$ and is $n \times n$. You will be asked to establish these results in the exercises that follow. Whenever idempotent matrices are referred to in this text, it is assumed they are also symmetric.

THEOREM 7.1.1. If $\mathbf{X}$ is p-variate normal with $\boldsymbol{\mu} = \mathbf{0}$, $\boldsymbol{\Sigma} = \mathbf{I}$, then $\mathbf{X}'\mathbf{A}\mathbf{X}$ is χ_r^2 if and only if $\mathbf{A}$ is idempotent of rank r.

Proof: The moment generating function for $\mathbf{X}'\mathbf{A}\mathbf{X}$ is

$$\begin{aligned} m_{\mathbf{X}'\mathbf{A}\mathbf{X}}(t) &= E[e^{t\mathbf{X}'\mathbf{A}\mathbf{X}}] \\ &= \int_{-\infty}^{\infty} \cdots \int_{-\infty}^{\infty} e^{t\mathbf{x}'\mathbf{A}\mathbf{x}} \frac{1}{(2\pi)^{p/2}} e^{-(\frac{1}{2})\mathbf{x}'\mathbf{x}}\, d\mathbf{x} \\ &= \frac{1}{|\mathbf{I} - 2t\mathbf{A}|^{\frac{1}{2}}} \int_{-\infty}^{\infty} \cdots \int_{-\infty}^{\infty} \frac{|\mathbf{I} - 2t\mathbf{A}|^{\frac{1}{2}}}{(2\pi)^{p/2}} e^{-(\frac{1}{2})\mathbf{x}'(\mathbf{I}-2t\mathbf{A})\mathbf{x}}\, d\mathbf{x} \\ &= \frac{1}{|\mathbf{I} - 2t\mathbf{A}|^{\frac{1}{2}}}. \end{aligned}$$

Since $\mathbf{A}$ is real and symmetric, there exists an orthogonal matrix $\mathbf{C}$ such that

$$\mathbf{C}'\mathbf{A}\mathbf{C} = \mathbf{D} = \text{Diag}(\lambda_1, \lambda_2, \ldots, \lambda_p)$$

where the λ_i's are the characteristic roots of $\mathbf{A}$. Since $\mathbf{C}$ is orthogonal $|\mathbf{C}| = \pm 1$ and

$$\begin{aligned} |\mathbf{I} - 2t\mathbf{A}| &= |\mathbf{C}'|\,|\mathbf{I} - 2t\mathbf{A}|\,|\mathbf{C}| \\ &= |\mathbf{C}'\mathbf{C} - 2t\mathbf{C}'\mathbf{A}\mathbf{C}| \\ &= |\mathbf{I} - 2t\mathbf{D}| = \prod_{i=1}^{p} (1 - 2t\lambda_i). \end{aligned}$$

Thus

$$m_{\mathbf{X}'\mathbf{A}\mathbf{X}}(t) = \frac{1}{\prod_{i=1}^{p} (1 - 2t\lambda_i)^{\frac{1}{2}}}.$$

If this is to equal $1/[(1 - 2t)^{r/2}]$, then exactly r of the λ_i's must be 1 and all the rest 0, since $\prod_{i=1}^{p}(1 - 2t\lambda_i)$ is a polynomial of degree p if all λ_i's $\neq 0$; furthermore, each nonzero λ_i must be 1 by equating like powers of t in $\prod_{i=1}^{r} (1 - 2t\lambda_i) = (1 - 2t)^r$, since this must hold for all real t in a continuous interval. Thus $\mathbf{A}$ must be idempotent of rank r for $\mathbf{X}'\mathbf{A}\mathbf{X}$ to be χ_r^2. Conversely, if $\mathbf{A}$ is idempotent of rank r, then

$$|\mathbf{I} - 2t\mathbf{A}| = (1 - 2t)^r$$

and $\mathbf{X}'\mathbf{A}\mathbf{X}$ is χ_r^2.

Since $\mathbf{I} - \mathbf{A}$ is also idempotent, and of rank $p - r$, note immediately then that if $\mathbf{X}'\mathbf{A}\mathbf{X}$ is χ^2_r, then $\mathbf{X}'(\mathbf{I} - \mathbf{A})\mathbf{X}$ is $\chi^2_{(p-r)}$. These two χ^2 random variables are also independent, as can be established by applying the following theorems.

THEOREM 7.1.2. If $\mathbf{X}$ is p-variate normal with $\boldsymbol{\mu} = \mathbf{0}$ and $\boldsymbol{\Sigma} = \mathbf{I}$, $\mathbf{Y} = \mathbf{H}\mathbf{X}$ and $\mathbf{Z} = \mathbf{M}\mathbf{X}$ are independent vectors if and only if $\mathbf{M}\mathbf{H}' = \mathbf{H}\mathbf{M}' = \mathbf{0}$. (The two vectors are called independent if every possible pair Y_i, Z_j are independent random variables.)

Proof: Clearly $\mathbf{Y}$ and $\mathbf{Z}$ are each multivariate normal vectors with $\mathbf{0}$ mean vectors and every pair Y_i, Z_j is jointly distributed. Then we need only show that $\text{Cov}(Y_i, Z_j) = 0$ for all i, j, and we have that all pairs are independent, that is $\mathbf{Y}$ and $\mathbf{Z}$ are independent vectors. All these covariances are the elements of

$$E[\mathbf{Y}\mathbf{Z}'] = E[\mathbf{H}\mathbf{X}\mathbf{X}'\mathbf{M}'] = \mathbf{H}\mathbf{M}',$$

and thus $\mathbf{Y}$ and $\mathbf{Z}$ are independent vectors if and only if $\mathbf{H}\mathbf{M}' = \mathbf{0}$ (or equivalently $(\mathbf{H}\mathbf{M}')' = \mathbf{M}\mathbf{H}' = \mathbf{0}$) and the proof is complete.

The proof of the following theorem will be left for the reader. Both parts follow quite easily from the given condition.

THEOREM 7.1.3. Each of (a) and (b) below are true if and only if $\mathbf{H}\mathbf{X}$ and $\mathbf{M}\mathbf{X}$ are independent, where $\mathbf{X}$ is p-variate normal, $\mathbf{0}$, $\mathbf{I}$.

(a) $\mathbf{X}'\mathbf{H}'\mathbf{H}\mathbf{X}$ and $\mathbf{X}'\mathbf{M}'\mathbf{M}\mathbf{X}$ are independent,
(b) $\mathbf{H}\mathbf{X}$ and $\mathbf{X}'\mathbf{M}'\mathbf{M}\mathbf{X}$ are independent.

Notice then, using this result with Theorem 7.1.2, $\mathbf{X}'\mathbf{H}'\mathbf{H}\mathbf{X}$ and $\mathbf{X}'\mathbf{M}'\mathbf{M}\mathbf{X}$ are independent if and only if $\mathbf{H}\mathbf{M}' = \mathbf{0}$, given $\mathbf{X}$ is multivariate normal with $\boldsymbol{\mu} = \mathbf{0}$ and $\boldsymbol{\Sigma} = \mathbf{I}$. The same is true for the independence of the linear form $\mathbf{H}\mathbf{X}$ and the quadratic form $\mathbf{X}'\mathbf{M}'\mathbf{M}\mathbf{X}$. Thus, if $\mathbf{X}'\mathbf{C}\mathbf{X}$ and $\mathbf{X}'\mathbf{D}\mathbf{X}$ each have idempotent matrices, then $\mathbf{C}'\mathbf{C} = \mathbf{C}^2 = \mathbf{C}$ and the necessary and sufficient condition for independence is $\mathbf{C}\mathbf{D} = 0$. (This is also true if they are not idempotent.) This result is very useful in showing independence in the analysis of variance.

In fact, if $\mathbf{X}$ is p-variate normal, $\mathbf{0}$, $\mathbf{I}$, $\mathbf{X}'\mathbf{X} = \sum_{i=1}^{p} X_i^2$ is the sum of squares of the components of $\mathbf{X}$; it is a χ^2_p random variable. Now if $\mathbf{A}$ is any $p \times p$ idempotent matrix of rank $r < p$, we have the obvious algebraic identity

$$\mathbf{X}'\mathbf{X} = \mathbf{X}'\mathbf{A}\mathbf{X} + \mathbf{X}'(\mathbf{I} - \mathbf{A})\mathbf{X};$$

$\mathbf{I} - \mathbf{A}$ is also idempotent, of rank $p - r$, and thus $\mathbf{X}'\mathbf{A}\mathbf{X}$ is χ^2_r, $\mathbf{X}'(\mathbf{I} - \mathbf{A})\mathbf{X}$ is χ^2_{p-r} and, since $\mathbf{A}(\mathbf{I} - \mathbf{A}) = \mathbf{0}$, $\mathbf{X}'\mathbf{A}\mathbf{X}$ and $\mathbf{X}'(\mathbf{I} - \mathbf{A})\mathbf{X}$ are independent. Thus, we have partitioned a χ^2_p random variable, the sum of squares of all the components of $\mathbf{X}$, into the sum of two other independently distributed χ^2 random variables. This partitioning of sums of squares, which we shall see can be looked at as a variance, into independent pieces is the basis of the term "*analysis of variance*"; the partitioning is done to analyze the variance.

For many applications, it is desirable to partition a variance into $k > 2$ pieces, which again may be independent χ^2 random variables. *Cochran's theorem*, which we shall now prove, shows a necessary and sufficient condition for determining this.

THEOREM 7.1.4. (*Cochran*). $\mathbf{X}$ is p-variate normal, $\mathbf{0}$, $\mathbf{I}$ and $\mathbf{X}'\mathbf{X}$ is χ^2_p. Assume $\mathbf{X}'\mathbf{X} = \sum_{i=1}^k \mathbf{X}'\mathbf{A}_i\mathbf{X}$ where $r(\mathbf{A}_i) = r_i$. Then the $\mathbf{X}'\mathbf{A}_i\mathbf{X}$ are independently distributed and $\mathbf{X}'\mathbf{A}_i\mathbf{X}$ is $\chi^2_{r_i}$ if and only if

$$\sum_{i=1}^{k} r_i = p.$$

Proof: First, let us assume $\mathbf{X}'\mathbf{X} = \sum_{i=1}^k \mathbf{X}'\mathbf{A}_i\mathbf{X}$ and that the $\mathbf{X}'\mathbf{A}_i\mathbf{X}$ are independent, and $\mathbf{X}'\mathbf{A}_i\mathbf{X}$ is $\chi^2_{r_i}$. Then the sum $\sum_{i=1}^k \mathbf{X}'\mathbf{A}_i\mathbf{X}$ must be $\chi^2_{\sum_{i=1}^k r_i}$; since $\mathbf{X}'\mathbf{X}$ is χ^2_p, we have $\sum_{i=1}^k r_i = p$. Now let us assume $\mathbf{X}'\mathbf{X} = \sum_{i=1}^k \mathbf{X}'\mathbf{A}_i\mathbf{X}$ and $\sum_{i=1}^k r_i = p$, where $r(\mathbf{A}_i) = r_i$. This matrix equation implies that

$$\sum_{i=1}^{k} \mathbf{A}_i = \mathbf{I}.$$

Then $\sum_{i=2}^k \mathbf{A}_i = \mathbf{I} - \mathbf{A}_1$, and we know

$$r\left(\sum_{i=2}^{k} \mathbf{A}_i\right) = r(\mathbf{I} - \mathbf{A}_1) \leq \sum_{i=2}^{k} r_i = p - r_1$$

since the rank of a sum is no larger than the sum of the ranks. Also, $\mathbf{I} = \mathbf{A}_1 + (\mathbf{I} - \mathbf{A}_1)$ so that $p \leq r_1 + r(\mathbf{I} - \mathbf{A}_1)$, and we must have $r(\mathbf{I} - \mathbf{A}_1) = p - r_1$. Since the rank of $\mathbf{A}_1$ is r_1, it must have exactly r_1 nonzero characteristic roots; that is, there are r_1 nonzero roots of $|\mathbf{A}_1 - \lambda\mathbf{I}| = 0$; since $r(\mathbf{I} - \mathbf{A}_1) = p - r_1$, the equation $|(\mathbf{I} - \mathbf{A}_1) - \lambda\mathbf{I}| = 0$ has $p - r_1$ nonzero roots, and r_1 zero roots. This last determinant can be written

$$|-\mathbf{A}_1 + (1 - \lambda)\mathbf{I}| = (-1)^p\, |\mathbf{A}_1 - (1 - \lambda)\mathbf{I}|,$$

and thus the roots of $\mathbf{I} - \mathbf{A}_1$ are the solutions of $|\mathbf{A}_1 - (1 - \lambda)\mathbf{I}| = 0$. Each root of this equation is one minus a characteristic root of $\mathbf{A}_1$; thus the r_1 zero roots of this equation correspond to the r_1 nonzero roots of $\mathbf{A}_1$. Therefore, $\mathbf{A}_1$ has exactly r_1 roots that are 1, and the rest are 0. We know then that $\mathbf{A}_1$ is idempotent and thus that $\mathbf{X}'\mathbf{A}_1\mathbf{X}$ is $\chi^2_{r_1}$, by Theorem 7.1.1. Exactly the same argument applied to $\mathbf{A}_2, \mathbf{A}_3, \ldots, \mathbf{A}_k$, in turn, will show that each of these is also idempotent; thus we have $\mathbf{X}'\mathbf{A}_i\mathbf{X}$ is $\chi^2_{r_i}$, $i = 1, 2, \ldots, k$ and then

$$m_{\mathbf{X}'\mathbf{A}_i\mathbf{X}}(t) = \frac{1}{(1-2t)^{r_i/2}}, \qquad i = 1, 2 \ldots, k.$$

We can easily see that the joint moment generating function of the $\mathbf{X}'\mathbf{A}_i\mathbf{X}$, $i = 1, 2, \ldots, k$, is

$$E[e^{t_1\mathbf{X}'\mathbf{A}_1\mathbf{X}+t_2\mathbf{X}'\mathbf{A}_2\mathbf{X}+\cdots+t_k\mathbf{X}'\mathbf{A}_k\mathbf{X}}] = E\left[e^{\mathbf{X}'\left(\sum_{i=1}^{k} t_i\mathbf{A}_i\right)\mathbf{X}}\right] = \frac{1}{\left|\mathbf{I} - 2\sum_{i=1}^{k} t_i\mathbf{A}_i\right|^{1/2}}$$

using the integral evaluated in the proof of Theorem 7.1.1. It also can be shown (see Exercise 11 below) that with our given conditions $\mathbf{A}_i\mathbf{A}_j = 0$ for all $i \neq j$. Granted this we have

$$\prod_{i=1}^{k} (\mathbf{I} - 2t_i\mathbf{A}_i) = \left(\mathbf{I} - 2\sum_{i=1}^{k} t_i\mathbf{A}_i\right)$$

and thus

$$\left|\mathbf{I} - 2\sum_{i=1}^{k} t_i\mathbf{A}_i\right| = \prod_{i=1}^{k} |\mathbf{I} - 2t_i\mathbf{A}_i|$$

$$= \prod_{i=1}^{k} (1 - 2t_i)^{r_i}$$

since each $\mathbf{A}_i$ is idempotent and we have established that the joint moment generating function of $\mathbf{X}'\mathbf{A}_i\mathbf{X}$ is the product of their marginal moment generating functions; thus the $\mathbf{X}'\mathbf{A}_i\mathbf{X}$ are independent χ^2 random variables as was to be established.

EXERCISE 7.1

1. If $\mathbf{A}$ is a symmetric matrix, show that $\mathbf{A}^2 = \mathbf{A}$ if and only if each characteristic root of $\mathbf{A}$ is 0 or 1. (Hint: There exists an orthogonal $\mathbf{C}$ such that $\mathbf{C}'\mathbf{A}\mathbf{C} = \mathbf{D} = \text{Diag}(\lambda_1, \ldots, \lambda_p)$ where the λ_i's are the characteristic roots. Clearly, if the λ_i's are 0 or 1, $\mathbf{D}^2 = \mathbf{D}$, i.e., $\mathbf{C}'\mathbf{A}\mathbf{C}\mathbf{C}'\mathbf{A}\mathbf{C} = \mathbf{C}'\mathbf{A}\mathbf{C}$.)

2. If $\mathbf{A}$ is idempotent, show that $\text{tr}(\mathbf{A}) = r(\mathbf{A})$; that is the rank is given by the sum of the diagonal elements.

3. If $\mathbf{A}$ is idempotent, show that $\mathbf{I} - \mathbf{A}$ is idempotent also.

4. Prove that the only idempotent matrix of full rank (non-singular) is $\mathbf{I}$.

5. Assume **X** is p-variate normal with mean vector **0** and covariance matrix **V**. Show that a necessary and sufficient condition that $\mathbf{X}'\mathbf{A}\mathbf{X}$ is χ_r^2, where $r = r(\mathbf{A})$ is $\mathbf{AVA} = \mathbf{A}$. (Equivalently, **AV** is idempotent.)

6. Prove Theorem 7.1.3.

7. Assume $X_1, X_2, \ldots, X_n$ is a random sample of a (scalar) standard normal random variable and define $\mathbf{X}' = (X_1, X_2, \ldots, X_n)$. Then the sample mean is

$$\bar{X} = \frac{1}{n}\sum X_i = \frac{1}{n}\mathbf{1}'\mathbf{X}$$

where $\mathbf{1}' = (1, 1, \ldots, 1)$. Let $\mathbf{J} = \mathbf{1}\mathbf{1}' = \|1\|$, the $n \times n$ matrix with 1 in every position. Verify that the sample variance is given by

$$S^2 = \frac{1}{n-1}\sum_{i=1}^{n}(X_i - \bar{X})^2 = \frac{1}{n-1}\mathbf{X}'\left(\mathbf{I} - \frac{1}{n}\mathbf{J}\right)\mathbf{X}$$

and use Theorems 7.1.2 and 7.1.3 to show that $(n - 1)S^2$ and $\bar{X}$ are independent.

8. Assume **X** is defined as in question 7 above and use Cochran's theorem to prove that $(n - 1)S^2$ and $n\bar{X}^2$ are independent χ^2 random variables.

9. Assume $X_1, X_2, \ldots, X_n$ is a random sample of a scalar normal random variable with mean 0 and variance σ^2, where n is an even number. Define

$$U = \frac{(X_1 - X_2)^2 + (X_3 - X_4)^2 + \cdots + (X_{n-1} - X_n)^2}{2\sigma^2}$$

and show that U is a χ_r^2 random variable. Evaluate r and use a linear function of U to define an unbiased estimator of σ^2. Why might the sample variance be preferred to this as an estimator of σ^2?

10. $X_1, X_2, \ldots, X_n$ are as defined in question 9, except the population mean is $\mu \neq 0$. By defining

$$\mathbf{Y}' = (X_1 - X_2, X_3 - X_4, \ldots, X_{n-1} - X_n)$$

write U as a quadratic form in **Y** and again verify U is a χ_r^2 random variable.

11. Assume $\mathbf{A}_1$, $\mathbf{A}_2$, $\mathbf{A}_3$ are each idempotent and $\sum \mathbf{A}_i = \mathbf{I}$. Show that $\mathbf{A}_1\mathbf{A}_2 = \mathbf{A}_1\mathbf{A}_3 = \mathbf{A}_2\mathbf{A}_3 = \mathbf{0}$. (Hint: Look at $\mathbf{A}_1 \sum \mathbf{A}_i = \mathbf{A}_1$, etc.) Use induction to generalize this result to k idempotent matrices $\mathbf{A}_1, \mathbf{A}_2, \ldots, \mathbf{A}_k$.

The following exercises extend the results of this section to a multivariate normal random variable with nonzero mean and arbitrary covariance matrix. Most of them are quite straightforward and can be proved with methods very similar to those used in proving the above theorems.

12. Assume **X** is p-variate normal with arbitrary $\boldsymbol{\mu}$ and (positive definite) **V**. Similar to Theorem 7.1.1, show that $\mathbf{X}'\mathbf{A}\mathbf{X}$ is $\chi_r'^2$ (noncentral χ^2 with r degrees of freedom where $r = r(\mathbf{A})$) with noncentrality $\boldsymbol{\mu}'\mathbf{A}\boldsymbol{\mu}$ if and only if $\mathbf{AVA} = \mathbf{A}$. (Hint: Use the nonsingular matrix **C** such that $\mathbf{CVC}' = \mathbf{I}$ to define $\mathbf{Y} = \mathbf{CX}$; then **Y** has mean $\mathbf{C}\boldsymbol{\mu}$ and covariance **I**.)

13. Assume $X_1, X_2, \ldots, X_n$ is a random sample of a scalar normal random variable with mean μ and variance σ^2. Define $\mathbf{X}' = (X_1, X_2, \ldots, X_n)$ as in question 7 above and use the result derived in question 10 to show that $(n-1)S^2/\sigma^2$ is a central χ^2_{n-1} random variable no matter what the value of μ. Also show that $(n\bar{X}^2)/\sigma^2$ is a noncentral χ_1^2 ($\chi_1'^2$) with noncentrality $(n\mu^2)/\sigma^2$.

14. Similar to Theorem 7.1.2, show that if $\mathbf{X}$ is p-variate normal, $\boldsymbol{\mu}$, $\mathbf{V}$, then $\mathbf{HX}$ and $\mathbf{MX}$ are independent if and only if $\mathbf{HVM}' = \mathbf{0}$.

15. Use the result of problem 11 and Theorem 7.1.3 to show that the two quadratic forms $\mathbf{X}'\mathbf{AX}$ and $\mathbf{X}'\mathbf{BX}$ are independent if and only if $\mathbf{AVB} = \mathbf{0}$.

16. Let $\mathbf{X}$ be defined as in question 10 and use the result of question 12 to show that S^2 and $\bar{X}^2$ are independent.

17. Extend Cochran's theorem to the case in which $\boldsymbol{\mu} \neq \mathbf{0}$, still assuming the covariance matrix is $\mathbf{I}$.

7.2 Least Squares Estimation

In many applied problems two or more associated variables are measured and, from the data, it is desired to estimate the relationship between them. Frequently the relationship can be phrased as follows: We have n measurements on $Y, x_1, x_2, \ldots, x_p$ and for each set of fixed values of $x_1, x_2, \ldots, x_p$, there is a population or distribution of values of Y, only one of which has been observed. Furthermore, for any given set of values $x_1, x_2, \ldots, x_p$

$$E[Y] = \beta_1 x_1 + \beta_2 x_2 + \cdots + \beta_p x_p$$

where $\beta_1, \beta_2, \ldots, \beta_p$ are unknown parameters; that is, there is a hyperplane in the space of $Y, x_1, x_2, \ldots, x_p$ and the mean value of Y lies on this hyperplane. The variance of the population of Y values is identical for each of the sets of values of $x_1, x_2, \ldots, x_p$. The first problem we shall discuss is the estimation of the unknown parameters from the sample data using the method of least squares.

Assume then that we have n sets of data $(Y_i, x_{i1}, x_{i2}, \ldots, x_{ip})$, $i = 1, 2, \ldots, n$, and define the $n \times 1$ vector $\mathbf{Y}' = (Y_1, Y_2, \ldots, Y_n)$, the $n \times p$ matrix

$$\mathbf{x} = \begin{pmatrix} x_{11} & x_{12} & \cdots & x_{1p} \\ x_{21} & x_{22} & \cdots & x_{2p} \\ \cdot & & & \\ \cdot & & & \\ \cdot & & & \\ x_{n1} & x_{n2} & \cdots & x_{np} \end{pmatrix}$$

and the $p \times 1$ vector $\boldsymbol{\beta}' = (\beta_1, \beta_2, \ldots, \beta_p)$. We assume that $E[\mathbf{Y}] = \mathbf{x}\boldsymbol{\beta}$, where the components of the vector $\boldsymbol{\beta}$ are unknown. We also assume that the components of the vector $\mathbf{Y}$ are independent, each with the same (unknown) variance σ^2. Thus, the covariance matrix of $\mathbf{Y}$ is $\sigma^2\mathbf{I}$. We would like to use the sample data to estimate the components of $\boldsymbol{\beta}$ and the scalar σ^2.

The *method of least squares* says we should choose that vector $\boldsymbol{\beta}$ which minimizes $Q = [\mathbf{Y} - \mathbf{x}\boldsymbol{\beta}]'[\mathbf{Y} - \mathbf{x}\boldsymbol{\beta}]$; notice that $\mathbf{Y} - \mathbf{x}\boldsymbol{\beta}$ is measuring the difference between the observed vector $\mathbf{Y}$ and possible linear combinations $\mathbf{x}\boldsymbol{\beta}$ of the columns of $\mathbf{x}$. Then Q has formed the sum of squares of differences between the components of $\mathbf{Y}$ and $\mathbf{x}\boldsymbol{\beta}$. By minimizing this sum of squares with respect to $\boldsymbol{\beta}$, we are choosing that value $\hat{\mathscr{B}}$ which makes $\mathbf{x}\hat{\mathscr{B}}$ as close to $\mathbf{Y}$ as possible, in a sense. Since we are minimizing a sum of squares, the name least squares is used to describe the method.

If we let $\mathbf{x}_1, \mathbf{x}_2, \ldots, \mathbf{x}_n$ represent the rows of the matrix $\mathbf{x}$ in partitioned form, thus

$$\mathbf{x} = \begin{pmatrix} \mathbf{x}_1 \\ \mathbf{x}_2 \\ \cdot \\ \cdot \\ \cdot \\ \mathbf{x}_n \end{pmatrix}, \quad \mathbf{Y} - \mathbf{x}\boldsymbol{\beta} = \begin{pmatrix} Y_1 - \mathbf{x}_1\boldsymbol{\beta} \\ Y_2 - \mathbf{x}_2\boldsymbol{\beta} \\ \cdot \\ \cdot \\ \cdot \\ Y_n - \mathbf{x}_n\boldsymbol{\beta} \end{pmatrix}$$

and we can write $Q = \sum_{i=1}^{n} (Y_i - \mathbf{x}_i\boldsymbol{\beta})^2$. Since Q is a continuous function of $\boldsymbol{\beta}$, we would expect to find the vector $\hat{\mathscr{B}}$ that minimizes Q by setting the partial derivatives, $\partial Q/\partial \beta_i$, $i = 1, 2, \ldots, p$, simultaneously equal to zero and solving the resulting equations. Every one of the terms $(Y_i - \mathbf{x}_i\boldsymbol{\beta})^2$, $i = 1, 2, \ldots, n$, involves β_1. Thus we would have

$$\frac{\partial Q}{\partial \beta_1} = -2\sum_{i=1}^{n}(Y_i - \mathbf{x}_i\boldsymbol{\beta})x_{i1} = -2\sum_{i=1}^{n} x_{i1}Y_i + 2\sum_{i=1}^{n} x_{i1}\mathbf{x}_i\boldsymbol{\beta};$$

similarly it is easy to see that, for $j = 1, 2, \ldots, p$, we have

$$\frac{\partial Q}{\partial \beta_j} = -2\sum_{i=1}^{n} x_{ij}Y_i + 2\sum_{i=1}^{n} x_{ij}\mathbf{x}_i\boldsymbol{\beta}.$$

Setting all these equations simultaneously equal to zero gives

$$\sum_{i=1}^{n} x_{ij}\mathbf{x}_i\hat{\mathscr{B}} = \sum_{i=1}^{n} x_{ij}Y_i, \qquad j = 1, 2, \ldots, p$$

as the equations to be solved for $\hat{\mathscr{B}}$. Note that in vector notation this is equivalent to

$$\mathbf{x}_j^{*\prime}\mathbf{x}\hat{\mathscr{B}} = \mathbf{x}_j^{*\prime}\mathbf{Y}, \qquad j = 1, 2, \ldots, p,$$

where $\mathbf{x}_j^*$, $j = 1, 2, \ldots, p$, are the column vectors of $\mathbf{x}$, and thus all p equations are simultaneously expressed in the matrix equation

$$\mathbf{x}'\mathbf{x}\hat{\mathscr{B}} = \mathbf{x}'\mathbf{Y}.$$

We shall now add the assumption that $\mathbf{x}'\mathbf{x}$ is nonsingular, that is, that the columns of $\mathbf{x}$ are linearly independent vectors. This assumption is essentially always satisfied in *regression models;* when we examine *experimental design models* in Section 7.4 this assumption cannot be made. There we shall investigate some of the consequences of $\mathbf{x}'\mathbf{x}$ being a singular matrix. With $\mathbf{x}'\mathbf{x}$ nonsingular, the unique least squares estimator of $\boldsymbol{\beta}$ is of course

$$\hat{\mathscr{B}} = (\mathbf{x}'\mathbf{x})^{-1}\mathbf{x}'\mathbf{Y}.$$

Let us first investigate the mean vector of $\hat{\mathscr{B}}$. We see that

$$E[\hat{\mathscr{B}}] = E[(\mathbf{x}'\mathbf{x})^{-1}\mathbf{x}'\mathbf{Y}] = (\mathbf{x}'\mathbf{x})^{-1}\mathbf{x}'E[\mathbf{Y}] = (\mathbf{x}'\mathbf{x})^{-1}\mathbf{x}'\mathbf{x}\boldsymbol{\beta} = \boldsymbol{\beta}$$

so $\hat{\mathscr{B}}$ is an unbiased estimator of $\boldsymbol{\beta}$. Its variance covariance matrix is

$$\begin{aligned} E[(\hat{\mathscr{B}} - \boldsymbol{\beta})(\hat{\mathscr{B}} - \boldsymbol{\beta})'] &= E[((\mathbf{x}'\mathbf{x})^{-1}\mathbf{x}'\mathbf{Y} - \boldsymbol{\beta})((\mathbf{x}'\mathbf{x})^{-1}\mathbf{x}'\mathbf{Y} - \boldsymbol{\beta})'] \\ &= E[(\mathbf{x}'\mathbf{x})^{-1}\mathbf{x}'(\mathbf{Y} - \mathbf{x}\boldsymbol{\beta})(\mathbf{Y} - \mathbf{x}\boldsymbol{\beta})'\mathbf{x}(\mathbf{x}'\mathbf{x})^{-1}] \\ &= (\mathbf{x}'\mathbf{x})^{-1}\mathbf{x}'(\sigma^2\mathbf{I})\mathbf{x}(\mathbf{x}'\mathbf{x})^{-1} \\ &= \sigma^2(\mathbf{x}'\mathbf{x})^{-1}, \end{aligned}$$

since we can factor the constants (functions of $\mathbf{x}$) out on both the left and right side of the expectation. Thus, the matrix $(\mathbf{x}'\mathbf{x})^{-1}$ allows us to evaluate $\hat{\mathscr{B}}$ and also gives the variances and covariances of $\hat{\mathscr{B}}$, apart from the scalar σ^2.

The method of least squares does not itself provide a rationale for estimating σ^2. However, recalling that in scalar examples a good estimator of σ^2 was based on the squares of the differences $Y_i - \bar{Y}$, we would be led to investigate the squares of the differences in the components of $\mathbf{Y}$ and of $\mathbf{x}\hat{\mathscr{B}}$, since the various components of $\mathbf{x}\hat{\mathscr{B}}$ are our least squares estimates of the corresponding mean values of the components of $\mathbf{Y}$. If $E[Y_i] = \mu$ for all i, then $\bar{Y}$ estimates $E[Y_i]$ for each i. Accordingly

$$\begin{aligned} E[(\mathbf{Y} - \mathbf{x}\hat{\mathscr{B}})'(\mathbf{Y} - \mathbf{x}\hat{\mathscr{B}})] &= E[(\mathbf{Y} - \mathbf{x}(\mathbf{x}'\mathbf{x})^{-1}\mathbf{x}'\mathbf{Y})'(\mathbf{Y} - \mathbf{x}(\mathbf{x}'\mathbf{x})^{-1}\mathbf{x}'\mathbf{Y})] \\ &= E[\mathbf{Y}'(\mathbf{I} - \mathbf{x}(\mathbf{x}'\mathbf{x})^{-1}\mathbf{x}')^2\mathbf{Y}] \\ &= E[\mathbf{Y}'(\mathbf{I} - \mathbf{x}(\mathbf{x}'\mathbf{x})^{-1}\mathbf{x}')\mathbf{Y}], \end{aligned}$$

since it is readily verified that $(\mathbf{I} - \mathbf{x}(\mathbf{x}'\mathbf{x})^{-1}\mathbf{x}')$ is an idempotent matrix. Furthermore, it is easily seen that $(\mathbf{I} - \mathbf{x}(\mathbf{x}'\mathbf{x})^{-1}\mathbf{x}')\mathbf{x} = \mathbf{0}$; thus we have changed nothing if we subtract $\mathbf{x}\boldsymbol{\beta}$ from each of the two $\mathbf{Y}$'s above, and we can write

$$E[(\mathbf{Y} - \mathbf{x}\hat{\mathscr{B}})'(\mathbf{Y} - \mathbf{x}\hat{\mathscr{B}})] = E[(\mathbf{Y} - \mathbf{x}\boldsymbol{\beta})'(\mathbf{I} - \mathbf{x}(\mathbf{x}'\mathbf{x})^{-1}\mathbf{x}')(\mathbf{Y} - \mathbf{x}\boldsymbol{\beta})].$$

Now $(\mathbf{Y} - \mathbf{x}\boldsymbol{\beta})'\mathbf{A}(\mathbf{Y} - \mathbf{x}\boldsymbol{\beta}) = \sum_{i=1}^{n} \sum_{j=1}^{n} a_{ij}(Y_i - \mathbf{x}_i\boldsymbol{\beta})(Y_j - \mathbf{x}_j\boldsymbol{\beta})$ and if we take the expectation of both sides, the expected value of each squared term $(Y_i - \mathbf{x}_i\boldsymbol{\beta})^2$ is σ^2 and the expected value of each cross-product is 0 since the variance covariance matrix of $\mathbf{Y}$ is $\sigma^2\mathbf{I}$. Thus

$$E[(\mathbf{Y} - \mathbf{x}\boldsymbol{\beta})'\mathbf{A}(\mathbf{Y} - \mathbf{x}\boldsymbol{\beta})] = \sigma^2 \sum_{i=1}^{n} a_{ii} = \sigma^2 \operatorname{tr}(\mathbf{A})$$

and we have

$$\begin{aligned} E[(\mathbf{Y} - \mathbf{x}\hat{\mathscr{B}})'(\mathbf{Y} - \mathbf{x}\hat{\mathscr{B}})] &= \sigma^2 \operatorname{tr}(\mathbf{I} - \mathbf{x}(\mathbf{x}'\mathbf{x})^{-1}\mathbf{x}') \\ &= \sigma^2[\operatorname{tr}(\mathbf{I}) - \operatorname{tr}(\mathbf{x}(\mathbf{x}'\mathbf{x})^{-1}\mathbf{x}')] \\ &= \sigma^2[n - \operatorname{tr}(\mathbf{x}'\mathbf{x}(\mathbf{x}'\mathbf{x})^{-1})] \\ &= \sigma^2[n - \operatorname{tr}(\mathbf{I})] \\ &= \sigma^2[n - p] \end{aligned}$$

since $\mathbf{x}'\mathbf{x}$ is $p \times p$ and the trace is unchanged by cyclic permutation in a product. We can define an unbiased estimator of σ^2, then by

$$S^2 = [1/(n - p)](\mathbf{Y} - \mathbf{x}\hat{\mathscr{B}})'(\mathbf{Y} - \mathbf{x}\hat{\mathscr{B}}).$$

(A handier computational formula is $S^2 = [1/(n - p)][\mathbf{Y}'\mathbf{Y} - \hat{\mathscr{B}}'\mathbf{x}'\mathbf{Y}]$.) Let us now examine the form of these estimators for a problem in *simple linear regression* with one independent variable x.

Example 7.2.1

Assume that eight Monterey pine trees, each about one foot tall, were selected to be as identical as possible. They were then transplanted to separate plots and irrigated individually during one year to simulate differing amounts of rainfall. At the end of one year, their achieved heights were measured. These quantities, together with the amounts of simulated rainfall, are presented in Table 7.2.1 below:

TABLE 7.2.1. Achieved Height (y_i) and Amount of Rainfall (x_i) for 8 Monterey Pines

y_i	19	22	25	31	33	39	44	45
x_i	10	14	18	22	26	30	34	38

We shall assume that the achieved height of any one of these trees at the end of the year is the observed value of a random variable Y with variance σ^2 and that the mean height for this random variable is $a + bx$, where x is the rainfall during the year; we also assume that the amounts of growth of the 8 trees are independent random variables. Thus, to put this example in terms of the general method just described, $n = 8$, $p = 2$, $\mathbf{Y}' = (Y_1, Y_2, \ldots, Y_8)$,

$$\mathbf{x}' = \begin{pmatrix} 1 & 1 & 1 & \cdots & 1 \\ x_1 & x_2 & x_3 & \cdots & x_8 \end{pmatrix}, \qquad \boldsymbol{\beta} = \begin{pmatrix} a \\ b \end{pmatrix},$$

and $x_1, x_2, \ldots, x_8$ have the values given in Table 7.2.1. Note then that

$$\mathbf{x}'\mathbf{x} = \begin{pmatrix} 8 & \sum x_i \\ \sum x_i & \sum x_i^2 \end{pmatrix}$$

and its inverse is

$$(\mathbf{x}'\mathbf{x})^{-1} = \frac{1}{8\sum (x_i - \bar{x})^2} \begin{pmatrix} \sum x_i^2 & -\sum x_i \\ -\sum x_i & 8 \end{pmatrix}.$$

We also find for this simple linear regression model,

$$\mathbf{x}'\mathbf{Y} = \begin{pmatrix} \sum Y_i \\ \sum x_i Y_i \end{pmatrix}$$

and thus

$$\mathscr{B} = \begin{pmatrix} \hat{A} \\ \hat{B} \end{pmatrix} = (\mathbf{x}'\mathbf{x})^{-1}\mathbf{x}'\mathbf{Y} = \frac{1}{8\sum_i (x_i - \bar{x})^2} \begin{pmatrix} \sum_i Y_i \sum_j x_j^2 - \sum_i x_i \sum_j x_j Y_j \\ 8\sum_i x_i Y_i - \sum_i x_i \sum_j Y_j \end{pmatrix}$$

$$= \begin{pmatrix} \bar{Y} - \bar{x}\dfrac{\sum (x_i - \bar{x})(Y_i - \bar{Y})}{\sum (x_i - \bar{x})^2} \\ \dfrac{\sum (x_i - \bar{x})(Y_i - \bar{Y})}{\sum (x_i - \bar{x})^2} \end{pmatrix}.$$

Furthermore, for this simple model,

$$\mathbf{Y}'\mathbf{Y} - \mathscr{B}'\mathbf{x}'\mathbf{Y} = \sum Y_i^2 - (\sum Y_i)(\bar{Y} - \hat{B}\bar{x}) - \hat{B}\sum x_i Y_i$$
$$= \sum (Y_i - \bar{Y})^2 - \hat{B}^2 \sum (x_i - \bar{x})^2.$$

For the observed data given in Table 7.2.1 we find $\sum x_i = 192$, $\sum y_i = 258$, $\sum x_i y_i = 6{,}864$, $\sum x_i^2 = 5{,}280$, $\sum y_i^2 = 9.002$, and thus $\sum (x_i - \bar{x})(y_i - \bar{y}) = 672$, $\sum (x_i - \bar{x})^2 = 672$, $\sum (y_i - \bar{y})^2 = 681\frac{1}{2}$, $\bar{x} = 24$, $\bar{y} = 32\frac{1}{4}$. The estimates of a and b then are $\hat{b} = 1$, $\hat{a} = 8\frac{1}{4}$ and the estimated value of σ^2 is $s^2 = (1/6)(9\frac{1}{2}) = 1.58$. The estimated variance of $\hat{A}$ is $(1.58)(.982) = 1.55$, the estimated variance of $\hat{B}$ is $(1.58)(.00149) = .0023$ and their estimated covariance is $(1.58)(-.0357) = -.056$.

The following theorem gives the main reason that least squares estimators are of value. Note that no specific assumption about the distribution is needed for Y, beyond the assumed mean and covariance structure.

THEOREM 7.2.1. (*Gauss-Markov*). Assume that $\mathbf{Y}$ is an $n \times 1$ random vector such that

(a) $E[\mathbf{Y}] = \mathbf{x}\boldsymbol{\beta}$, where $\mathbf{x}$ is an $n \times p$ matrix of known constants and $\boldsymbol{\beta}$ is a $p \times 1$ vector of unknown constants.

(b) $E[(\mathbf{Y} - \mathbf{x}\boldsymbol{\beta})(\mathbf{Y} - \mathbf{x}\boldsymbol{\beta})'] = \sigma^2\mathbf{I}$.

Then the *best linear unbiased estimators* of $\beta_1, \beta_2, \ldots, \beta_p$ are the components of $\hat{\mathscr{B}} = (\mathbf{x}'\mathbf{x})^{-1}\mathbf{x}'\mathbf{Y}$.

Proof: Clearly, p linear functions of the components of $\mathbf{Y}$ can generally be represented by

$$\mathscr{B}^* = \mathbf{aY} + \mathbf{b}$$

where $\mathbf{a}$ is an arbitrary $p \times n$ constant matrix and $\mathbf{b}$ is an arbitrary $p \times 1$ constant vector. Then,

$$E[\mathscr{B}^*] = \mathbf{a}\mathbf{x}\boldsymbol{\beta} + \mathbf{b};$$

thus for $\mathscr{B}^*$ to be unbiased, no matter what the true value of $\boldsymbol{\beta}$, we require $\mathbf{b} = \mathbf{0}$ and $\mathbf{ax} = \mathbf{I}$. Furthermore, for any given $\mathbf{a}$ we can find a $p \times n$ matrix $\mathbf{c}$ such that

$$\mathbf{a} = (\mathbf{x}'\mathbf{x})^{-1}\mathbf{x}' + \mathbf{c},$$

and thus we can write our general linear estimator as $\mathscr{B}^* = [(\mathbf{x}'\mathbf{x})^{-1}\mathbf{x}' + \mathbf{c}]\mathbf{Y}$; then

$$E[\mathscr{B}^*] = [(\mathbf{x}'\mathbf{x})^{-1}\mathbf{x}' + \mathbf{c}]\mathbf{x}\boldsymbol{\beta} = \boldsymbol{\beta} + \mathbf{c}\mathbf{x}\boldsymbol{\beta}$$

and if $\mathscr{B}^*$ is to be unbiased we require $\mathbf{c}\mathbf{x}\boldsymbol{\beta} = \mathbf{0}$, no matter what the unknown value of $\boldsymbol{\beta}$. This can only be satisfied if $\mathbf{cx} = 0$. Thus our general estimator now is represented by

$$\mathscr{B}^* = [(\mathbf{x}'\mathbf{x})^{-1}\mathbf{x}' + \mathbf{c}]\mathbf{Y}$$

where $\mathbf{c}$ must satisfy $\mathbf{cx} = \mathbf{0}$. The variance covariance matrix of $\mathscr{B}^*$ is

$$\begin{aligned} E[(\mathscr{B}^* - \boldsymbol{\beta})(\mathscr{B}^* - \boldsymbol{\beta})'] &= E[\{[(\mathbf{x}'\mathbf{x})^{-1}\mathbf{x}' + \mathbf{c}]\mathbf{Y} - \boldsymbol{\beta}\} \\ &\qquad \times \{[(\mathbf{x}'\mathbf{x})^{-1}\mathbf{x}' + \mathbf{c}]\mathbf{Y} - \boldsymbol{\beta}\}'] \\ &= [(\mathbf{x}'\mathbf{x})^{-1}\mathbf{x}' + \mathbf{c}]E[(\mathbf{Y} - \mathbf{x}\boldsymbol{\beta})(\mathbf{Y} - \mathbf{x}\boldsymbol{\beta})'] \\ &\qquad \times [(\mathbf{x}'\mathbf{x})^{-1}\mathbf{x}' + \mathbf{c}]' \\ &= [(\mathbf{x}'\mathbf{x})^{-1}\mathbf{x}' + \mathbf{c}][\sigma^2\mathbf{I}][(\mathbf{x}'\mathbf{x})^{-1}\mathbf{x}' + \mathbf{c}]' \\ &= \sigma^2[(\mathbf{x}'\mathbf{x})^{-1} + \mathbf{cc}'] \end{aligned}$$

since $\mathbf{cx} = \mathbf{x'c'} = \mathbf{0}$. Thus, the variance of the ith component of $\mathscr{B}^*$ is σ^2 times the ith diagonal element of $(\mathbf{x'x})^{-1}$ plus the ith diagonal element of $\mathbf{cc'}$. For any given problem of course, σ^2 and $(\mathbf{x'x})^{-1}$ are fixed and the only things we can vary are the elements of $\mathbf{cc'}$. Note that the ith diagonal of $\mathbf{cc'}$ is $\sum_{j=1}^{n} c_{ij}^2$, where c_{ij}, $j = 1, 2, \ldots, n$, are the elements of the ith row of $\mathbf{c}$. Clearly the smallest possible value of $\sum_{j=1}^{n} c_{ij}^2$ is zero and $\sum c_{ij}^2 = 0$ implies that we must have $c_{ij} = 0$ for each j. Thus, to simultaneously minimize the variances of all the components of $\mathscr{B}^*$ we must take $\mathbf{c} = \mathbf{0}$; this is consistent with the unbiasedness requirement that $\mathbf{cx} = \mathbf{0}$, and the best linear unbiased estimators of the components of $\boldsymbol{\beta}$ are given by $\mathscr{B}^* = [(\mathbf{x'x})^{-1}\mathbf{x'} + 0]\mathbf{Y} = \hat{\mathscr{B}}$.

This theorem is very powerful, and it is remarkable that the actual distribution of the components of $\mathbf{Y}$ does not have to be specified. Their joint distribution could be the multivariate normal, or they could be jointly gamma distributed, or any other distribution that can be made consistent with assumptions (a) and (b) stated in the theorem. Then, no matter what the specifics of the distribution, we know that we cannot find linear unbiased estimators of the β_j's that have a smaller variance than the components of $\hat{\mathscr{B}}$. As we shall see in the following section, if we assume that $\mathbf{Y}$ is multivariate normal, the maximum likelihood estimators of $\boldsymbol{\beta}$ are also given by $\hat{\mathscr{B}}$ and thus these estimators have good properties from two distinct points of view.

EXERCISE 7.2

1. Assume we have a random sample of n pairs, (Y_i, x_i), $i = 1, 2, \ldots, n$, the Y_i's are independent, the x_i's are given constants and

$$\left.\begin{aligned} E[Y_i] &= bx_i \\ \mathrm{Var}(Y_i) &= \sigma^2 \end{aligned}\right\} \qquad i = 1, 2, \ldots, n.$$

Find the best linear unbiased estimate of b. How would you estimate σ^2?

2. For the case of simple linear regression, as discussed in Example 7.2.1, show that the least squares line $\hat{Y} = \hat{A} + \hat{B}x$ passes through the point with coordinates $(\bar{x}, \bar{Y})$. Generalize this result to the case of p independent variables where $x_1 \equiv 1$ (thus we have a constant term); that is $E[Y_i] = \beta_1 + \beta_2 x_{i2} + \cdots + \beta_p x_{ip}$. Show that the point with coordinates $\bar{Y}, \bar{x}_2, \ldots, \bar{x}_p$ lies on the least squares hyperplane

$$\hat{Y} = \hat{B}_1 + \hat{B}_2 x_2 + \cdots + \hat{B}_p x_p.$$

(Hint: Define $\mathbf{1} = (1, 1, \ldots, 1)$ and examine $\mathbf{1x}\hat{\mathscr{B}} = \mathbf{1}\hat{\mathbf{Y}}$.)

3. If $\mathbf{Y}$ is a random vector and $\mathbf{a}$ is a constant matrix, show that

(a) $E[\mathbf{aYY'a'}] = \mathbf{a}E[\mathbf{YY'}]\mathbf{a'}$
(b) $E[\mathbf{Y'aY}] = \sigma^2 \,\mathrm{tr}(\mathbf{a})$, given $E[\mathbf{YY'}] = \sigma^2\mathbf{I}$.

4. If $\mathbf{x}$ is $n \times p$ of rank p, show that $\mathbf{x}(\mathbf{x}'\mathbf{x})^{-1}\mathbf{x}'$ and $\mathbf{I} - \mathbf{x}(\mathbf{x}'\mathbf{x})^{-1}\mathbf{x}'$ are each idempotent and that their product is $\mathbf{0}$.

5. Verify the identities:

$$(\mathbf{Y} - \mathbf{x}\hat{\mathscr{B}})'(\mathbf{Y} - \mathbf{x}\hat{\mathscr{B}}) = \mathbf{Y}'\mathbf{Y} - \hat{\mathscr{B}}'\mathbf{x}'\mathbf{x}\hat{\mathscr{B}} = \mathbf{Y}'\mathbf{Y} - \hat{\mathscr{B}}'\mathbf{x}'\mathbf{Y}.$$

6. Generalizc the Gauss-Markov theorem to the case in which $V(\mathbf{Y}) = \sigma^2\mathbf{w}$, where $\mathbf{w}$ is known and positive definite, and show that the best linear unbiased estimator of $\boldsymbol{\beta}$ is $\hat{\mathscr{B}} = (\mathbf{x}'\mathbf{w}^{-1}\mathbf{x})^{-1}\mathbf{x}'\mathbf{w}^{-1}\mathbf{Y}$. (Hint: There exists a real non-singular matrix $\mathbf{p}$ such that $\mathbf{w} = \mathbf{p}\mathbf{p}'$; define $\mathbf{Z} = \mathbf{p}^{-1}\mathbf{Y}$.)

7. With all the assumptions made in the Gauss-Markov theorem, prove that the best linear unbiased estimator of $\mathbf{t}'\boldsymbol{\beta}$, where $\mathbf{t}' = (t_1, t_2, \ldots, t_p)$ are any given constants, is $\mathbf{t}'\hat{\mathscr{B}}$. Note that the variance of $\mathbf{t}'\hat{\mathscr{B}}$ will involve the covariances as well as the variances of the components of $\hat{\mathscr{B}}$.

8. Using the result derived in problem 6, assume $Y_1, Y_2, \ldots, Y_n$ are independent, $x_1, x_2, \ldots, x_n$ are known positive constants and $E[Y_i] = bx_i$, $i = 1, 2, \ldots, n$. Derive the best linear unbiased estimator for b if

(a) $V(Y_i) = \sigma^2 x_i, \qquad i = 1, 2, \ldots, n.$
(b) $V(Y_i) = \sigma^2 x_i^2, \qquad i = 1, 2, \ldots, n.$

9. n different people counted the traffic on the same street, at about the same time of day, on separate independent occasions. Unfortunately, they did not all observe the traffic for the same length of time. Person i observed it for x_i minutes, $i = 1, 2, \ldots, n$. Assume that the traffic on this road passes the observation point like events in a Poisson process, with a rate of λ cars per minute. Thus, letting Y_i be the number of cars to pass the observation point when person i was there, we have Y_i is a Poisson random variable with parameter λx_i and

$$E[Y_i] = \lambda x_i, \qquad i = 1, 2, \ldots, n$$

$$V[Y_i] = \lambda x_i, \qquad i = 1, 2, \ldots, n.$$

What is the best linear unbiased estimator for λ?

10. For the situation described in question 9, a reasonable heuristic estimator for λ could be constructed by reasoning as follows:

$$E\left[\frac{Y_i}{\sqrt{x_i}}\right] = \lambda\sqrt{x_i}, \qquad V\left[\frac{Y_i}{\sqrt{{}_i x}}\right] = \lambda;$$

then

$$\frac{\sum_i \frac{Y_i}{\sqrt{x_i}}}{\sum \sqrt{x_i}}$$

is an unbiased estimator for λ. Compute the variance of this estimator and show that it is larger than that of the estimator derived in 9.

11. Assume $E[\mathbf{Y}] = \mathbf{x\beta}$, $V[\mathbf{Y}] = \sigma^2\mathbf{w}$ as in exercise 6. If the columns of $\mathbf{x}$ are characteristic vectors of $\mathbf{x}$, show that the best linear unbiased estimator for $\boldsymbol{\beta}$ is

$$\hat{\mathscr{B}} = (\mathbf{x}'\mathbf{x})^{-1}\mathbf{x}'\mathbf{Y}.$$

12. Assume that y is a function of x, $y = f(x)$, but that the functional form f is unknown. Assuming that f is such that it can be expanded in a Taylor series about the origin, we can write

$$y = f(x) = f(0) + xf'(0) + \frac{x^2}{2}f''(0) + r$$

where r is the remainder; this could also be written

$$y = \beta_0 + \beta_1 x + \beta_2 x^2 + e$$

where $\beta_0 = f(0)$, $\beta_1 = f'(0)$, $\beta_2 = (1/2)f''(0)$, $e = r$, with β_0, β_1, β_2 unknown. Assuming that e (or r) may act like a "random" error, we could then apply the techniques of this chapter to estimate β_0, β_1, and β_2. Thus $\hat{\beta}_0 + \hat{\beta}_1 x + \hat{\beta}_2 x^2$ then would serve as an approximation to the unknown function $f(x)$.

(a) As the number of terms included in the expansion increases, the better the approximation. Why is this so?

(b) Generalize this approach. Assume that $y = f(x_1, x_2)$ and again expand in a Taylor series.

7.3 Normality and Maximum Likelihood

We saw in the last section that we can derive estimators of the components of $\boldsymbol{\beta}$, when $E[\mathbf{Y}] = \mathbf{x\beta}$, even when we make essentially no assumptions about the distribution of $\mathbf{Y}$. In many cases it is reasonable to assume that $\mathbf{Y}$ is a multivariate normal vector. We can then employ maximum likelihood to estimate the parameters in the distribution of $\mathbf{Y}$ and we can derive interval estimates of the parameters and test hypotheses about their values.

We shall assume that $\mathbf{Y}$ is a multivariate normal vector with mean vector $\mathbf{x\beta}$ and covariance matrix $\sigma^2\mathbf{I}$, where $\mathbf{x}$ is a known $n \times p$ matrix of rank p, $\boldsymbol{\beta}$ is a $p \times 1$ vector of unknown constants and σ^2 is an unknown scalar. The likelihood function of the sample then is

$$L(\boldsymbol{\beta}, \sigma^2) = \frac{1}{(2\pi\sigma^2)^{n/2}} e^{-(1/2\sigma^2)(\mathbf{y}-\mathbf{x\beta})'(\mathbf{y}-\mathbf{x\beta})}$$

and $\log L = -(n/2)\log 2\pi - (n/2)\log \sigma^2 - (1/2\sigma^2)(\mathbf{y} - \mathbf{x\beta})'(\mathbf{y} - \mathbf{x\beta})$. Since L (and $\log L$) is a continuous function of the unknown parameters, we would expect to find the values of $\boldsymbol{\beta}$ and σ^2 that maximize L by setting the partial derivatives of $\log L$ simultaneously equal to zero. Thus

$$\frac{\partial \log L}{\partial \sigma^2} = -\frac{n}{2\sigma^2} + \frac{1}{2(\sigma^2)^2}(\mathbf{y} - \mathbf{x\beta})'(\mathbf{y} - \mathbf{x\beta});$$

note that if we assume σ^2 is constant, $\log L = k - (1/2\sigma^2)Q$, where $Q = (\mathbf{y} - \mathbf{x}\boldsymbol{\beta})'(\mathbf{y} - \mathbf{x}\boldsymbol{\beta})$ as in Section 7.2. Thus, all the partial derivates of $\log L$ with respect to the components of $\boldsymbol{\beta}$ can be simultaneously displayed in the matrix equation

$$-\frac{1}{2\sigma^2}(-2\mathbf{x}'\mathbf{y} + 2\mathbf{x}'\mathbf{x}\boldsymbol{\beta}).$$

We see, then, that if we set all these partial derivatives simultaneously equal to zero, the solutions are

$$\hat{\mathscr{B}} = (\mathbf{x}'\mathbf{x})^{-1}\mathbf{x}'\mathbf{y}$$

$$\hat{\sigma}^2 = \frac{1}{n}(\mathbf{y} - \mathbf{x}\hat{\mathscr{B}})'(\mathbf{y} - \mathbf{x}\hat{\mathscr{B}}),$$

that is, $\hat{\mathscr{B}}$ is the same as the least squares estimator. Then, we know immediately that $\hat{\mathscr{B}}$ is an unbiased estimator for $\boldsymbol{\beta}$ but $\hat{\sigma}^2$ is biased; $S^2 = [1/(n - p)](\mathbf{Y} - \mathbf{x}\hat{\mathscr{B}})'(\mathbf{Y} - \mathbf{x}\hat{\mathscr{B}})$ is an unbiased estimator for σ^2. But, with the normality assumption for $\mathbf{Y}$ we can say much more than that; since $\hat{\mathscr{B}}$ is a linear function of the elements of $\mathbf{Y}$, $\hat{\mathscr{B}}$ has a multivariate normal distribution with mean vector $\boldsymbol{\beta}$ and covariance matrix $\sigma^2(\mathbf{x}'\mathbf{x})^{-1}$. Since

$$\begin{aligned}(n - p)S^2 &= (\mathbf{Y} - \mathbf{x}\hat{\mathscr{B}})'(\mathbf{Y} - \mathbf{x}\hat{\mathscr{B}})\\ &= \mathbf{Y}'(\mathbf{I} - \mathbf{x}(\mathbf{x}'\mathbf{x})^{-1}\mathbf{x}')\mathbf{Y}\\ &= (\mathbf{Y} - \mathbf{x}\boldsymbol{\beta})'(\mathbf{I} - \mathbf{x}(\mathbf{x}'\mathbf{x})^{-1}\mathbf{x}')(\mathbf{Y} - \mathbf{x}\boldsymbol{\beta})\end{aligned}$$

we know from Theorem 7.1.1 that $[(n - p)S^2]/\sigma^2$ is a χ^2 random variable with $r(\mathbf{I} - \mathbf{x}(\mathbf{x}'\mathbf{x})^{-1}\mathbf{x}') = \mathrm{tr}(\mathbf{I} - \mathbf{x}(\mathbf{x}'\mathbf{x})^{-1}\mathbf{x}') = n - p$ degrees of freedom, because $\mathbf{I} - \mathbf{x}(\mathbf{x}'\mathbf{x})^{-1}\mathbf{x}'$ is an idempotent matrix. Furthermore, from Theorems 7.1.2 and 7.1.3, $\hat{\mathscr{B}}$ and S^2 are independent since $(\mathbf{x}'\mathbf{x})^{-1}\mathbf{x}'(\mathbf{I} - \mathbf{x}(\mathbf{x}'\mathbf{x})^{-1}\mathbf{x}') = \mathbf{0}$. Some of the consequences of these results are illustrated in the following example.

Example 7.3.1

Let us make the assumption, for the data discussed in Example 7.2.1, that the achieved height of Monterey pine trees at the end of one year is a normal random variable with mean $a + bx$, where x is the amount of rainfall, and variance σ^2. Then we would know immediately that the maximum likelihood or least squares estimators $\hat{A}$ and $\hat{B}$, of a and b, are normal random variables, that $[(8 - 2)S^2]/\sigma^2 = 6S^2/\sigma^2$ (for the values used in the example) is a χ^2_6 random variable and that $\hat{A}$ and S^2 are independent, as are $\hat{B}$ and S^2. Also, the variance of $\hat{A}$ is $\sigma^2 c_{11}$ and the variance of $\hat{B}$ is $\sigma^2 c_{22}$, where

$$\mathbf{C} = \begin{pmatrix} c_{11} & c_{12} \\ c_{21} & c_{22} \end{pmatrix} = (\mathbf{x}'\mathbf{x})^{-1}.$$

Thus

$$\frac{\hat{A} - a}{\sigma\sqrt{c_{11}}}\bigg/\frac{S}{\sigma} = \frac{\hat{A} - a}{S\sqrt{c_{11}}} \quad \text{and} \quad \frac{\hat{B} - b}{\sigma\sqrt{c_{22}}}\bigg/\frac{S}{\sigma} = \frac{\hat{B} - b}{S\sqrt{c_{22}}}$$

each have a t distribution with $n - p = 6$ degrees of freedom. (They are not independent. Why?) We can then use these t distributions to place confidence intervals about a and b individually and to test hypotheses about a and about b. For the given sample data in Example 7.2.1, $(\hat{a} - t_6(.95)s\sqrt{c_{11}}, \hat{a} + t_6(.95)s\sqrt{c_{11}}) = (8\frac{1}{4} - (1.943)(1.58)\sqrt{.982}, 8\frac{1}{4} + (1.943)(1.58)\sqrt{.982}) = (5.21, 11.29)$ is an observed 90% confidence interval for a; similarly $(\hat{b} - t_6(.99)sc_{22}, \hat{b} + t_6(.99)sc_{22}) = (1 - (3.143)(1.58)\sqrt{.00149}, 1 + (3.143)(1.58) \times \sqrt{.00149}) = (.81, 1.19)$ is the observed 98% confidence interval for b. If we want to estimate the average height of a tree of this type in a location which will receive $x = 20$ inches of rain in the year, our point estimate of this average would be $\hat{a} + \hat{b}(20) = 28\frac{1}{4}$ inches, based on the sample data. Furthermore, $\hat{A} + \hat{B}x$ has variance $V(\hat{A}) + x^2V(\hat{B}) + 2x\,\mathrm{Cov}(\hat{A}, \hat{B}) = \sigma^2(c_{11} + x^2c_{22} + 2xc_{12})$. Thus the estimated variance of $\hat{A} + \hat{B}x$, with $x = 20$, is $(1.58)[.982 + 400(.00149) + 40(-.0357)] = .237$ and a 95% confidence interval for $a + 20b$ is $(28\frac{1}{4} - (2.447)\sqrt{.237}, 28\frac{1}{4} + (2.447) \times \sqrt{.237}) = (27.06, 29.44)$. We can also test hypotheses about the values of a, of b or of $a + bx$ in the obvious way using the t random variables just discussed.

In Example 7.3.1 we discussed several scalar techniques useful in examining a, or b, or σ^2 separately. In many practical problems it is desirable to be able to test hypotheses that make statements about several parameters simultaneously. For example, if we assume that the mean value of Y, given values for $x_2, x_3, \ldots, x_p$ is

$$E[Y] = \beta_1 + \beta_2 x_2 + \cdots + \beta_p x_p,$$

we might very well want to test the hypothesis that the mean value of Y is the same for all values of $x_2, x_3, \ldots, x_p$, i.e., there is no relation between Y and the independent variables. This is obviously equivalent to testing H_0: $\beta_2 = 0, \beta_3 = 0, \ldots, \beta_p = 0$ versus the alternative that at least one of these β_i's is not zero. Note that the value of β_1 has remained unspecified. We shall now examine the generalized likelihood ratio test criterion for hypotheses of this sort.

It will be recalled that a generalized likelihood ratio test consists of maximizing the likelihood function with the parameters restricted to lying in ω_0, the values specified by H_0; then letting $L(\hat{\omega}_0)$ represent this achieved maximum of the likelihood, we form the ratio $\lambda = (L(\hat{\omega}_0))/(L(\hat{\Omega}))$, where $L(\hat{\Omega})$ is the maximum of the likelihood, allowing the parameters to take on values specified by either H_0 or H_1. We reject H_0 if $\lambda < A$, where A is chosen to make our probability of type I error equal to α.

We shall make the following assumptions: $\mathbf{Y}$ is n-variate normal with mean $\mathbf{x}\boldsymbol{\beta}$ and variance covariance matrix $\sigma^2\mathbf{I}$ where $\mathbf{x}$ is an $n \times p$ matrix of known constants of rank p, $\boldsymbol{\beta}$ is a $p \times 1$ vector of unknown constants and σ^2 is an unknown scalar. To allow for the possibility of testing hypotheses about some components of $\boldsymbol{\beta}$ and leaving the others unspecified, we shall partition $\mathbf{x} = (\mathbf{x}_1, \mathbf{x}_2)$ and $\boldsymbol{\beta}' = (\boldsymbol{\beta}_1', \boldsymbol{\beta}_2')$, where $\mathbf{x}_1$ is $n \times s$, $\mathbf{x}_2$ is $n \times (p - s)$, $\boldsymbol{\beta}_1$ is $s \times 1$ and $\boldsymbol{\beta}_2$ is $(p - s) \times 1$. We can allow $0 \leq s \leq p - 1$; if $s = 0$ then we are testing a hypothesis about all the components of $\boldsymbol{\beta}$. Note then that $\mathbf{x}\boldsymbol{\beta} = \mathbf{x}_1\boldsymbol{\beta}_1 + \mathbf{x}_2\boldsymbol{\beta}_2$. We shall examine the likelihood ratio test of H_0: $\boldsymbol{\beta}_2 = \boldsymbol{\beta}_2^0$ versus H_1: $\boldsymbol{\beta}_2 \neq \boldsymbol{\beta}_2^0$, where $\boldsymbol{\beta}_2^0$ is any vector of specified constants; note that σ^2 and the values of $\boldsymbol{\beta}_1$ have remained unspecified by H_0. Thus, $\omega_0 = \{\boldsymbol{\beta}_2 = \boldsymbol{\beta}_2^0, \ \boldsymbol{\beta}_1, \ \sigma^2 \text{ unspecified}\}$, $\omega_1 = \{\boldsymbol{\beta}_2 \neq \boldsymbol{\beta}_2^0, \ \boldsymbol{\beta}_1, \ \sigma^2 \text{ unspecified}\}$ and $\Omega = \omega_0 \cup \omega_1$ is the full parameter space. The maximizing values for the parameters in Ω are the maximum likelihood estimates:

$$\hat{\mathscr{B}} = (\mathbf{x}'\mathbf{x})^{-1}\mathbf{x}'\mathbf{y}, \qquad \hat{\sigma}^2 = \frac{1}{n}(\mathbf{y} - \mathbf{x}\hat{\mathscr{B}})'(\mathbf{y} - \mathbf{x}\hat{\mathscr{B}}),$$

and the maximum value of L then is

$$\begin{aligned} L(\hat{\Omega}) &= \frac{1}{(2\pi)^{n/2}}\left[\frac{n}{(\mathbf{y} - \mathbf{x}\hat{\mathscr{B}})'(\mathbf{y} - \mathbf{x}\hat{\mathscr{B}})}\right]^{n/2} e^{-\frac{1}{2}\left(\frac{n}{(\mathbf{y}-\mathbf{x}\hat{\mathscr{B}})'(\mathbf{y}-\mathbf{x}\hat{\mathscr{B}})}\right)(\mathbf{y}-\mathbf{x}\hat{\mathscr{B}})'(\mathbf{y}-\mathbf{x}\hat{\mathscr{B}})} \\ &= \frac{e^{-(n/2)}}{(2\pi)^{n/2}}\left[\frac{n}{(\mathbf{y} - \mathbf{x}\hat{\mathscr{B}})'(\mathbf{y} - \mathbf{x}\hat{\mathscr{B}})}\right]^{n/2}. \end{aligned}$$

To maximize the likelihood function with the parameters restricted by ω_0, note that the likelihood is

$$\begin{aligned} L(\boldsymbol{\beta}_1, \boldsymbol{\beta}_2^0, \sigma^2) &= \frac{1}{(2\pi\sigma^2)^{n/2}} e^{-(1/2\sigma^2)(\mathbf{y}-\mathbf{x}_1\boldsymbol{\beta}_1-\mathbf{x}_2\boldsymbol{\beta}_2^0)'(\mathbf{y}-\mathbf{x}_1\boldsymbol{\beta}_1-\mathbf{x}_2\boldsymbol{\beta}_2^0)} \\ &= \frac{1}{(2\pi\sigma^2)^{n/2}} e^{-(1/2\sigma^2)(\mathbf{z}-\mathbf{x}_1\boldsymbol{\beta}_1)'(\mathbf{z}-\mathbf{x}_1\boldsymbol{\beta}_1)} \end{aligned}$$

where $\mathbf{z} = \mathbf{y} - \mathbf{x}_2\boldsymbol{\beta}_2^0$ and the only free variables are $\boldsymbol{\beta}_1$ and σ^2. We can see immediately then that this is the same maximization problem we just solved, with $\mathbf{y}$ replaced by $\mathbf{z}$, $\mathbf{x}$ by $\mathbf{x}_1$ and $\boldsymbol{\beta}$ by $\boldsymbol{\beta}_1$. Thus the maximizing values are

$$\hat{\mathscr{B}}_1 = (\mathbf{x}_1'\mathbf{x}_1)^{-1}\mathbf{x}_1'\mathbf{z} = (\mathbf{x}_1'\mathbf{x}_1)^{-1}\mathbf{x}_1'(\mathbf{y} - \mathbf{x}_2\boldsymbol{\beta}_2^0)$$

$$\hat{\sigma}^2 = \frac{1}{n}(\mathbf{z} - \mathbf{x}_1\hat{\mathscr{B}}_1)'(\mathbf{z} - \mathbf{x}_1\hat{\mathscr{B}}_1)$$

$$= \frac{1}{n}(\mathbf{y} - \mathbf{x}_1\hat{\mathscr{B}}_1 - \mathbf{x}_2\boldsymbol{\beta}_2^0)'(\mathbf{y} - \mathbf{x}_1\hat{\mathscr{B}}_1 - \mathbf{x}_2\boldsymbol{\beta}_2^0)$$

and the restricted maximum value of the likelihood function is

$$L(\hat{\omega}_0) = \frac{e^{-(n/2)}}{(2\pi)^{n/2}}\left[\frac{n}{(\mathbf{z} - \mathbf{x}_1\hat{\mathscr{B}}_1)'(\mathbf{z} - \mathbf{x}_1\hat{\mathscr{B}}_1)}\right]^{n/2}.$$

The likelihood ratio test criterion is the ratio of these two:

$$\lambda = \frac{L(\hat{\omega}_0)}{L(\hat{\Omega})} = \left[\frac{(\mathbf{y} - \mathbf{x}\hat{\mathscr{B}})'(\mathbf{y} - \mathbf{x}\hat{\mathscr{B}})}{(\mathbf{z} - \mathbf{x}_1\hat{\mathscr{B}}_1)'(\mathbf{z} - \mathbf{x}_1\hat{\mathscr{B}}_1)}\right]^{n/2}.$$

It is important to realize that the first s components of $\hat{\mathscr{B}}$ are not necessarily the same as $\hat{\mathscr{B}}_1$, since $\hat{\mathscr{B}}$ is the solution of

$$\mathbf{x}'\mathbf{x}\hat{\mathscr{B}} = \mathbf{x}'\mathbf{y}$$

and $\hat{\mathscr{B}}_1$ is the solution to

$$\mathbf{x}_1'\mathbf{x}_1\hat{\mathscr{B}}_1 = \mathbf{x}_1'(\mathbf{y} - \mathbf{x}_2\boldsymbol{\beta}_2^0);$$

these sets of equations are quite different, and thus their solutions are not identical.

We know that $L(\hat{\Omega}) \geq L(\hat{\omega}_0)$ and thus

$$(\mathbf{y} - \mathbf{x}\hat{\mathscr{B}})'(\mathbf{y} - \mathbf{x}\hat{\mathscr{B}}) \leq (\mathbf{z} - \mathbf{x}\hat{\mathscr{B}}_1)'(\mathbf{z} - \mathbf{x}\hat{\mathscr{B}}_1),$$

so let us define

$$Q_0 = (\mathbf{y} - \mathbf{x}\hat{\mathscr{B}})'(\mathbf{y} - \mathbf{x}\hat{\mathscr{B}})$$

$$Q_0 + Q_1 = (\mathbf{z} - \mathbf{x}_1\hat{\mathscr{B}}_1)'(\mathbf{z} - \mathbf{x}_1\hat{\mathscr{B}}_1).$$

The likelihood ratio test criterion can be written

$$\lambda = \left[\frac{Q_0}{Q_0 + Q_1}\right]^{n/2} = \left[\frac{1}{1 + Q_1/Q_0}\right]^{n/2};$$

as we shall see below, the difference of the two quadratic forms above, Q_1, is also a quadratic form. A very important identity, which we derived in Section 7.2, is

$$(*) \qquad Q_0 = (\mathbf{y} - \mathbf{x}\hat{\mathscr{B}})'(\mathbf{y} - \mathbf{x}\hat{\mathscr{B}}) = \mathbf{y}'(\mathbf{I} - \mathbf{x}(\mathbf{x}'\mathbf{x})^{-1}\mathbf{x}')\mathbf{y};$$

also

$$(\mathbf{I} - \mathbf{x}(\mathbf{x}'\mathbf{x})^{-1}\mathbf{x}')\mathbf{x} = 0$$

and thus, writing $\mathbf{x} = (\mathbf{x}_1, \mathbf{x}_2)$ we know that

$$(\mathbf{I} - \mathbf{x}(\mathbf{x}'\mathbf{x})^{-1}\mathbf{x}')\mathbf{x}_1 = \mathbf{0},$$

and

$$(\mathbf{I} - \mathbf{x}(\mathbf{x}'\mathbf{x})^{-1}\mathbf{x}')\mathbf{x}_2 = \mathbf{0},$$

i.e.,

$$\mathbf{x}_1 = \mathbf{x}(\mathbf{x}'\mathbf{x})^{-1}\mathbf{x}'\mathbf{x}_1$$
$$\mathbf{x}_2 = \mathbf{x}(\mathbf{x}'\mathbf{x})^{-1}\mathbf{x}'\mathbf{x}_2.$$

Then Q_0 can equally well be written

$$\begin{aligned} Q_0 &= (\mathbf{y} - \mathbf{x}_2\boldsymbol{\beta}_2^0)'(\mathbf{I} - \mathbf{x}(\mathbf{x}'\mathbf{x})^{-1}\mathbf{x}')(\mathbf{y} - \mathbf{x}_2\boldsymbol{\beta}_2^0) \\ &= \mathbf{z}'(\mathbf{I} - \mathbf{x}(\mathbf{x}'\mathbf{x})^{-1}\mathbf{x}')\mathbf{z} = \mathbf{z}'\mathbf{A}_0\mathbf{z} \end{aligned}$$

as a quadratic form in $\mathbf{z}$. Applying the identity (*) to $Q_0 + Q_1$, we have

$$\begin{aligned} Q_0 + Q_1 &= (\mathbf{z} - \mathbf{x}_1\hat{\mathscr{B}}_1)'(\mathbf{z} - \mathbf{x}_1\hat{\mathscr{B}}_1) \\ &= \mathbf{z}'(\mathbf{I} - \mathbf{x}_1(\mathbf{x}_1'\mathbf{x}_1)^{-1}\mathbf{x}_1')\mathbf{z}, \end{aligned}$$

again a quadratic form in $\mathbf{z}$. Their difference, Q_1, then is

$$\begin{aligned} Q_1 &= \mathbf{z}'(\mathbf{I} - \mathbf{x}_1(\mathbf{x}_1'\mathbf{x}_1)^{-1}\mathbf{x}_1')\mathbf{z} - \mathbf{z}'(\mathbf{I} - \mathbf{x}(\mathbf{x}'\mathbf{x})^{-1}\mathbf{x}')\mathbf{z} \\ &= \mathbf{z}'(\mathbf{x}(\mathbf{x}'\mathbf{x})^{-1}\mathbf{x}' - \mathbf{x}_1(\mathbf{x}_1'\mathbf{x}_1)^{-1}\mathbf{x}_1')\mathbf{z} = \mathbf{z}'\mathbf{A}_1\mathbf{z}. \end{aligned}$$

We saw earlier that $\mathbf{A}_0$ is idempotent of rank $n - p$ and thus that Q_0 is χ^2_{n-p}, no matter what the true vector $\boldsymbol{\beta}$ may be. It is readily verified that $\mathbf{A}_1$ is also idempotent and thus

$$\begin{aligned} r(\mathbf{A}_1) = \operatorname{tr}(\mathbf{A}_1) &= \operatorname{tr}(\mathbf{x}(\mathbf{x}'\mathbf{x})^{-1}\mathbf{x}') - \operatorname{tr}(\mathbf{x}_1(\mathbf{x}_1'\mathbf{x}_1)^{-1}\mathbf{x}_1') \\ &= \operatorname{tr}(\mathbf{x}'\mathbf{x}(\mathbf{x}'\mathbf{x})^{-1}) - \operatorname{tr}(\mathbf{x}_1'\mathbf{x}_1(\mathbf{x}_1'\mathbf{x}_1)^{-1}) \\ &= p - s, \end{aligned}$$

so Q_1 is χ'^2_{p-s} with noncentrality

$$\frac{[\mathbf{x}_2(\boldsymbol{\beta}_2 - \boldsymbol{\beta}_2^0)]'\mathbf{A}_1[\mathbf{x}_2(\boldsymbol{\beta}_2 - \boldsymbol{\beta}_2^0)]}{\sigma^2} = \frac{(\boldsymbol{\beta}_2 - \boldsymbol{\beta}_2^0)'\mathbf{x}_2'(\mathbf{I} - \mathbf{x}_1(\mathbf{x}_1'\mathbf{x}_1)^{-1}\mathbf{x}_1')\mathbf{x}_2(\boldsymbol{\beta}_2 - \boldsymbol{\beta}_2^0)}{\sigma^2}.$$

The matrix $\mathbf{x}_2'(\mathbf{I} - \mathbf{x}_1(\mathbf{x}_1'\mathbf{x}_1)^{-1}\mathbf{x}_1')\mathbf{x}_2$ is positive definite; thus the noncentrality parameter for Q_1 is 0, and Q_1 is central χ^2_{p-s}, if and only if $\boldsymbol{\beta}_2 - \boldsymbol{\beta}_2^0 = 0$ (H_0 is true). We can also see, by performing the multiplication, that $\mathbf{A}_0\mathbf{A}_1 = \mathbf{0}$, and, by Theorems 7.1.2 and 7.1.3, Q_0 and Q_1 are independent random variables.

The ratio

$$\frac{Q_1}{p - s}\Big/\frac{Q_0}{n - p}$$

then has a central $F_{p-s,n-p}$ distribution if and only if H_0 is true. Letting

$$F = \frac{Q_1}{p-s}\bigg/\frac{Q_0}{n-p}$$

we can write the likelihood ratio as

$$\lambda = \left(\frac{1}{1 + \dfrac{(p-s)}{(n-p)}F}\right)^{n/2}$$

where $\lambda < A$ implies $F > F_0$. Thus, the likelihood ratio test criterion says we should accept H_0: $\boldsymbol{\beta}_2 = \boldsymbol{\beta}_2^0$, rather than H_1: $\boldsymbol{\beta}_2 \neq \boldsymbol{\beta}_2^0$, if $F \leq F_{p,n-p}(1 - \alpha)$ and we will then have a probability of type I error equal to α.

Generally the quantities necessary to test such a hypothesis as H_0: $\boldsymbol{\beta}_2 = \boldsymbol{\beta}_2^0$ versus H_1: $\boldsymbol{\beta}_2 \neq \boldsymbol{\beta}_2^0$ are displayed in tabular form, called an *analysis of variance*. If in fact H_0 is true, then the sum of squares $(\mathbf{Y} - \mathbf{x}_2\boldsymbol{\beta}_2^0)'(\mathbf{Y} - \mathbf{x}_2\boldsymbol{\beta}_2^0)$ remains unexplained by the hypothesis. This unexplained variation or *variance* is analyzed (subdivided) in the analysis of variance; two of the parts in the analysis are Q_0 and Q_1, defined above, which are used to make the F test of H_0. Note the following obvious algebraic identity:

$$\mathbf{Z}'\mathbf{Z} = \mathbf{Z}'\mathbf{A}_0\mathbf{Z} + \mathbf{Z}'\mathbf{A}_1\mathbf{Z} + \mathbf{Z}'(\mathbf{I} - \mathbf{A}_0 - \mathbf{A}_1)\mathbf{Z}$$

where $\mathbf{Z} = \mathbf{Y} - \mathbf{x}_2\boldsymbol{\beta}_2^0$, $\mathbf{A}_0$ and $\mathbf{A}_1$ are as defined above. The tabular presentation of this analysis of variance is given in Table 7.3.1. The F test of H_0 is then simply the ratio of the mean square due to $\boldsymbol{\beta}_2$, eliminating $\boldsymbol{\beta}_1$, to the mean square for error. If this ratio exceeds $F_{p-s,n-p}(1 - \alpha)$ we reject H_0. The column labelled degrees of freedom is simply the number of degrees of freedom in the χ^2 distribution (central or noncentral) of the corresponding sum of squares, divided by σ^2, and the corresponding mean (average) square is the sum of squares divided by the degrees of freedom. The following example illustrates the methodology for testing such a hypothesis.

TABLE 7.3.1. Analysis of Variance for Testing H_0: $\boldsymbol{\beta}_2 = \boldsymbol{\beta}_2^0$ versus H_1: $\boldsymbol{\beta}_2 \neq \boldsymbol{\beta}_2^0$

Source of variation	Degrees of freedom	Sum of squares	Mean squares
Due to β_1, assuming $\beta_2 = \beta_2^0$	s	$\mathbf{Z}'(\mathbf{I} - \mathbf{A}_0 - \mathbf{A}_1)\mathbf{Z} = Q_2$	
Due to β_2, eliminating β_1	$p - s$	$\mathbf{Z}'\mathbf{A}_1\mathbf{Z} = Q_1$	$Q_1/(p - s)$
Unexplained (error)	$n - p$	$\mathbf{Z}'\mathbf{A}_0\mathbf{Z} = Q_0$	$Q_0/(n - p)$
Total	n	$\mathbf{Z}'\mathbf{Z}$	

Example 7.3.2

Table 7.3.2 presents the weights in ounces (y_i), ages in months (x_{i1}), and lengths in inches (x_{i2}) from a random sample of ten babies. It would seem reasonable to assume that the mean value of the weight depends linearly on both the length and the age of the baby, at least for limited ranges on these variables. Thus, letting Y_i be the weight of the ith baby, we might assume that

$$E[Y_i] = a_0 + a_1 x_{i1} + a_2 x_{i2}.$$

TABLE 7.3.2. Weights, Ages, and Lengths of Ten Babies

y = Weight (oz.)	x_1 = Age (mo.)	x_2 = Length (in.)
118	1	$21\frac{1}{2}$
162	5	$26\frac{1}{2}$
137	2	22
110	1	20
211	6	29
195	4	26
191	3	25
125	2	20
227	6	27
187	3	$22\frac{1}{2}$

Let us illustrate the above material by testing the hypothesis H_0: $a_1 = 0$, $a_2 = 0$ versus the alternative H_1: $a_1 \neq 0$ or $a_2 \neq 0$.

We must first estimate all three parameters a_0, a_1, and a_2. The equations to be solved are

$$\begin{aligned} 10\hat{a}_0 + 33\hat{a}_1 + 239.5\hat{a}_2 &= 1{,}663 \\ 33\hat{a}_0 + 141\hat{a}_1 + 840.5\hat{a}_2 &= 6{,}104 \\ 239.5\hat{a}_0 + 840.5\hat{a}_1 + 5825.75\hat{a}_2 &= 40{,}844.5 \end{aligned}$$

and the solutions are $\hat{a}_0 = 15.2081$, $\hat{a}_1 = 11.8984$, and $\hat{a}_2 = 4.6692$. Then $Q_0 = \mathbf{y}'\mathbf{y} - \hat{\mathscr{B}}'\mathbf{x}'\mathbf{y} = 292{,}187 - 288{,}630.0443 = 3{,}556.9567$. To determine Q_1 we assume $a_1 = a_2 = 0$ and the equation determining $\hat{a}_0$ is

$$10\hat{a}_0 = 1663$$

and the residual variance for this reduced model then is $Q_1 + Q_0 = \mathbf{y}'\mathbf{y} - \hat{\mathscr{B}}_1'\mathbf{x}_1'\mathbf{y} = 15{,}630.1$. Note that the solution for $\hat{a}_0$ differs considerably in the two cases, as was mentioned earlier. We also have $\mathbf{y}'\mathbf{y} = 292{,}187$ for the total sum of squares and $\hat{\mathscr{B}}_1'\mathbf{x}_1'\mathbf{y} = 166.3(1663) = 276{,}556.9$ for the sum of squares due to $\hat{a}_0$, assuming $a_1 = a_2 = 0$. These quantities are given in the following analysis of variance, Table 7.3.3. We see that $(Q_1/2)/(Q_0/7) = 11.88 > F_{2,7}(.99) = 9.55$ so we would reject H_0 with $\alpha = .01$.

TABLE 7.3.3. Analysis of Variance for Testing H_0: $a_1 = a_2 = 0$

Source	Degrees of freedom	Sums of squares	Mean squares
$\hat{a}_0$, assuming $a_1 = a_2 = 0$	1	276,556.9	
$\hat{a}_1$, $\hat{a}_2$, corrected for a_0	2	12,073.1	6,036.57
Residual	7	3,557.0	508.14
Total	10	292,187	

EXERCISE 7.3

1. For the data discussed in Example 7.3.1, would you accept H_0: $b = .75$ versus H_1: $b > .75$ with $\alpha = .05$? Would you accept the hypothesis that the achieved mean height of trees receiving 25 inches of rain is at least 35 inches, with $\alpha = .1$?

2. In Example 7.3.1, how would you place a confidence interval about σ^2? For the given data, would you accept H_0: $\sigma \leq 2$ versus H_1: $\sigma > 2$, with $\alpha = .01$?

3. With the assumption of normality made in this section, show that the estimators $\hat{\mathscr{B}}$ and $\hat{\sigma}^2$ are sufficient statistics for $\boldsymbol{\beta}$ and σ^2.

4. Verify that $\mathbf{A}_1 = \mathbf{x}(\mathbf{x}'\mathbf{x})^{-1}\mathbf{x}' - \mathbf{x}_1(\mathbf{x}_1'\mathbf{x}_1)^{-1}\mathbf{x}_1'$ is idempotent.

5. Show that $\mathbf{x}_2'(\mathbf{I} - \mathbf{x}_1(\mathbf{x}_1'\mathbf{x}_1)^{-1}\mathbf{x}_1')\mathbf{x}_2$ is positive definite, where $\mathbf{x} = (\mathbf{x}_1, \mathbf{x}_2)$ is $n \times p$ of rank p.

6. Use the data presented in Table 7.3.2 to test H_0: $a_1 = 14$, $a_2 = 8$ versus H_1: $a_1 \neq 14$ or $a_2 \neq 8$, with $\alpha = .05$.

7. Many other types of hypotheses may be of interest in regression problems. With the same notation used in this section, how would you test H_0: $\beta_{p-1} = \beta_p$, versus H_1: $\beta_{p-1} \neq \beta_p$, where their common value remains unspecified? (Hint: If H_0 is true, $\beta_0 + \beta_1 X_1 + \cdots + \beta_{p-1}X_{p-1} + \beta_p X_p = \beta_0 + \beta_1 X_1 + \cdots + \beta_{p-1}(X_{p-1} + X_p)$.)

8. Use the data from Table 7.3.2 to test H_0: $a_1 = a_2$ versus H_1: $a_1 \neq a_2$, where their common value remains unspecified, with $\alpha = .05$.

9. It will be recalled from Sections 1.4 and 7.1 that if $\mathbf{X}$ is p-variate normal with mean vector $\boldsymbol{\mu}$ and covariance $\boldsymbol{\Sigma}$, then $(\mathbf{X} - \boldsymbol{\mu})'\boldsymbol{\Sigma}^{-1}(\mathbf{X} - \boldsymbol{\mu})$ has the χ_p^2 distribution. We saw in this section, with the normality assumption for $\mathbf{Y}$, that $\hat{\mathscr{B}}$ is p-variate normal with vector of means $\boldsymbol{\beta}$ and covariance $\sigma^2(\mathbf{x}'\mathbf{x})^{-1}$; thus

$$(\hat{\mathscr{B}} - \boldsymbol{\beta})'[\sigma^2(\mathbf{x}'\mathbf{x})^{-1}]^{-1}(\hat{\mathscr{B}} - \boldsymbol{\beta}) = \frac{(\hat{\mathscr{B}} - \boldsymbol{\beta})'\mathbf{x}'\mathbf{x}(\hat{\mathscr{B}} - \boldsymbol{\beta})}{\sigma^2}$$

has the χ_p^2 distribution. $Q_0/\sigma^2 = [(\mathbf{Y} - \mathbf{x}\hat{\mathscr{B}})'(\mathbf{Y} - \mathbf{x}\hat{\mathscr{B}})]/\sigma^2$ has the χ_{n-p}^2 distribution and is independent of $\hat{\mathscr{B}}$. Derive a $100(1 - \alpha)\%$ confidence region for $\boldsymbol{\beta}$. (Hint: Consider a ratio which has an F distribution.)

10. Apply the method in question 9 to derive a confidence region for a and b, where $Y_1, Y_2, \ldots, Y_n$ are independent normal random variables, each with variance σ^2 and $E[Y_i] = a + bx_i$, $i = 1, 2, \ldots, n$. If this region is projected onto the a and b axes, respectively, are the resulting intervals the same as the individual confidence intervals described earlier?

7.4 Experimental Design Models

Most of the commonly used experimental design models were first adopted in agricultural experimentation. They are very special cases of the models we have discussed in Sections 7.2 and 7.3, ones in which the independent variables are binary variables, taking on only the values 0 and 1, and in the most commonly used notation, $\mathbf{x}'\mathbf{x}$ is singular. Let us begin by examining a specific example, called a *completely randomized design*. Suppose we wanted to compare the effects of three different types of insect spray on the yield of a crop. We have fifteen separate plots of ground available, in the same location, each of which is of a size that can be individually sprayed. Each plot is to receive exactly one spray.

We plant the same crop on each plot, in the same way, and then randomly allocate the sprays to the plots, with the restriction that each spray will be used five times. After the crop is harvested, we then will have available the yield from each of the fifteen plots. Let Y_{ij} denote the yield from the jth plot which received the ith spray, $i = 1, 2, 3,\quad j = 1, 2, 3, 4, 5$. The usual experimental design model then assumes

$$E[Y_{ij}] = \mu + \tau_i, \qquad i = 1, 2, 3, \qquad j = 1, 2, \ldots, 5,$$

that the Y_{ij}'s are independent and normal and each has the same variance σ^2. The parameter μ represents the mean yield with no spray and τ_i, which may be positive or negative, is the change in mean yield caused by the ith spray. Thus, we would assume that there is a conceptual normal population of yields for each spray, that the variances of the populations are equal and that possibly the means of the populations are different (τ_1, τ_2, and τ_3 may be different). Notice that these assumptions can be represented by our regression models of the last two sections. Let $\mathbf{Y}$ be the 15×1 vector whose first 5 components are the 5 yields for spray 1, the next 5 are for spray 2 and the final 5 are those for spray 3. Let $\mathbf{x}$ be the 15×4 matrix, each of whose first 5 rows are (1, 1, 0, 0), whose next 5 are each (1, 0, 1, 0) and whose final 5 are each (1, 0, 0, 1). Defining $\boldsymbol{\beta}' = (\mu, \tau_1, \tau_2, \tau_3)$ we then have $E[\mathbf{Y}] = \mathbf{x}\boldsymbol{\beta}$ and $\mathbf{Y}$ is multivariate normal with covariance matrix $\sigma^2\mathbf{I}$. Written in this way, we can easily recognize this as being the sort of linear model we have already studied. The major concern in such models, of course, is to estimate and compare the treatment effects. Thus we would like to test H_0: $\tau_1 = \tau_2 = \tau_3$, where their common value remains unspecified.

Rather than work out the details for this specific example, let us adopt a more general notation for a *completely randomized design*. Suppose we want to compare k treatments (k different populations) and that each population is normal with variance σ^2. The mean of the ith population is $\mu + \tau_i$, $i = 1, 2, \ldots, k$. Assume we will have available a random sample of n_i observations from the ith population, $i = 1, 2, \ldots, k$. Let $\mathbf{y}$ be the $[\sum_{i=1}^k n_i] \times 1$ observed vector whose first n_1 components are the values from population 1, the next n_2 are those from population 2, etc. Similarly $\mathbf{x}$ is the $(\sum n_i) \times k$ matrix whose first n_1 rows are each $(1, 1, 0, 0, \ldots, 0)$, whose next n_2 rows are each $(1, 0, 1, 0, \ldots, 0)$, etc., and $\boldsymbol{\beta}' = (\mu, \tau_1, \tau_2, \ldots, \tau_k)$. We can easily see that in this case the matrix $\mathbf{x}$ has only k different, linearly independent rows and thus $r(\mathbf{x}'\mathbf{x}) = r(\mathbf{x}) = k$; but $\mathbf{x}'\mathbf{x}$ is $(k + 1) \times (k + 1)$ and is thus a singular matrix. In such a case the normal equations

$$\mathbf{x}'\mathbf{x}\hat{\mathscr{B}} = \mathbf{x}'\mathbf{y}$$

have not a single unique solution, but an infinite number of solutions. The reason this happens can be easily seen by examining the parameter structure we have used to represent the mean values of the different populations. The mean of population i is $\mu + \tau_i$; thus we can certainly expect to estimate the quantities $\mu + \tau_1, \mu + \tau_2, \ldots, \mu + \tau_k$ but we do not have enough information to separately estimate $\mu, \tau_1, \ldots, \tau_k$. This redundancy in the parameters causes the singularity of $\mathbf{x}'\mathbf{x}$.

We need not be concerned with the fact that there are an infinite number of solutions to the normal equations, because the things of major interest (namely the estimation of σ^2 and of comparisons of the treatment effects) remain invariant for all solutions. For example, we have seen that the maximum likelihood estimate of σ^2 is a constant times $\mathbf{y}'\mathbf{y} - \hat{\mathscr{B}}\mathbf{x}'\mathbf{y}$, where $\mathbf{x}'\mathbf{x}\hat{\mathscr{B}} = \mathbf{x}'\mathbf{y}$. But with the current model there are many vectors $\hat{\mathscr{B}}$ satisfying $\mathbf{x}'\mathbf{x}\hat{\mathscr{B}} = \mathbf{x}'\mathbf{y}$. Suppose both $\hat{\mathscr{B}}_1$ and $\hat{\mathscr{B}}_2$ satisfy the normal equations; then

$$\begin{aligned}\hat{\mathscr{B}}_1'\mathbf{x}'\mathbf{y} &= \hat{\mathscr{B}}_1'\mathbf{x}'\mathbf{x}\hat{\mathscr{B}}_2 \\ &= \mathbf{y}'\mathbf{x}'\hat{\mathscr{B}}_2 \\ &= \hat{\mathscr{B}}_2'\mathbf{x}'\mathbf{y},\end{aligned}$$

since the transpose of a scalar is itself. Thus, $\mathbf{y}'\mathbf{y} - \hat{\mathscr{B}}'\mathbf{x}'\mathbf{y}$ will have the same value no matter which solution we use to the normal equations. Note as well that, if $\boldsymbol{\eta}$ is any given constant vector, $\boldsymbol{\eta}'\mathbf{x}'\mathbf{y} = \boldsymbol{\eta}'\mathbf{x}'\mathbf{x}\hat{\mathscr{B}}_1 = \boldsymbol{\lambda}'\hat{\mathscr{B}}_1 = \boldsymbol{\eta}'\mathbf{x}'\mathbf{x}\hat{\mathscr{B}}_2 = \boldsymbol{\lambda}'\hat{\mathscr{B}}_2$, where $\mathbf{x}'\mathbf{x}\boldsymbol{\eta} = \boldsymbol{\lambda}$; that is, no matter what solution we take to the normal equations, $\boldsymbol{\lambda}'\hat{\mathscr{B}}$ is invariant so long as there is a vector $\boldsymbol{\eta}$ such that $\mathbf{x}'\mathbf{x}\boldsymbol{\eta} = \boldsymbol{\lambda}$. This requirement, of course, says $\boldsymbol{\lambda}$ must be a linear combination of the columns of $\mathbf{x}'\mathbf{x}$ or, equivalently, a linear combination of the rows of $\mathbf{x}$ itself since the columns (or rows) of $\mathbf{x}'\mathbf{x}$ and the rows of $\mathbf{x}$ span the same space. The linear combination $\boldsymbol{\lambda}'\boldsymbol{\beta}$ is said to be an *estimable function* if $\boldsymbol{\lambda}$ is a linear combination of the rows of $\mathbf{x}$. Since, in the case under discussion, $r(\mathbf{x}) = k$, we can clearly find exactly k linearly independent $\boldsymbol{\lambda}$'s, each of which is a linear combination of the rows of $\mathbf{x}$; these give rise to *linearly independent estimable functions.*

Let us now return to our completely randomized design. The normal equations are

$$\begin{pmatrix} n & n_1 & n_2 & \cdots & n_k \\ n_1 & n_1 & 0 & \cdots & 0 \\ n_2 & 0 & n_2 & \cdots & 0 \\ \cdot & \cdot & \cdot & & \\ \cdot & \cdot & \cdot & & \\ \cdot & \cdot & \cdot & & \\ n_k & 0 & 0 & \cdots & n_k \end{pmatrix} \begin{pmatrix} \hat{\mu} \\ \hat{\tau}_1 \\ \hat{\tau}_2 \\ \cdot \\ \cdot \\ \cdot \\ \hat{\tau}_k \end{pmatrix} = \begin{pmatrix} \sum_i \sum_j y_{ij} \\ \sum_j y_{1j} \\ \sum_j y_{2j} \\ \cdot \\ \cdot \\ \cdot \\ \sum_j y_{kj} \end{pmatrix}.$$

From the last k equations we clearly have $\hat{\mu} + \hat{\tau}_i = (1/n_i)\sum_j y_{ij} = \bar{y}_i$, the mean of the sample values for the ith treatment, and thus $\hat{\tau}_i = \bar{y}_i - \hat{\mu}$. The value of $\hat{\mu}$ is arbitrary and, no matter what value it is assigned, we have a complete solution by taking $\hat{\tau}_i = \bar{y}_i - \hat{\mu}$, $i = 1, 2, \ldots, k$. Notice that the difference between any two treatment effects, $\tau_i - \tau_j$, is estimable (by examining the rows of $\mathbf{x}$) and that the estimate is $\hat{\tau}_i - \hat{\tau}_j = (\bar{y}_i - \hat{\mu}) - (\bar{y}_j - \hat{\mu}) = \bar{y}_i - \bar{y}_j$, no matter what value is assigned to $\hat{\mu}$. In fact, any *comparison* of the treatment effects ($\sum_i c_i\tau_i$, where $\sum_i c_i = 0$) is estimable; the estimate of $\sum_i c_i\tau_i$ is $\sum_i c_i\bar{y}_i$.

It is also of interest to test the hypothesis that all the treatments are equal in their effects. To test H_0: $\tau_1 = \tau_2 = \cdots = \tau_t$ versus the alternative that H_0 is not true, we must first derive the minimum value of $(\mathbf{y} - \mathbf{x}\boldsymbol{\beta})'(\mathbf{y} - \mathbf{x}\boldsymbol{\beta})$, where the parameters are completely unrestricted, as discussed in the last section. This of course occurs with $\boldsymbol{\beta} = \hat{\mathscr{B}}$, where $\hat{\mathscr{B}}$ is any solution to the normal equations and the minimum value is

$$Q_0 = \mathbf{y}'\mathbf{y} - \hat{\mathscr{B}}'\mathbf{x}'\mathbf{y} = \sum_i \sum_j y_{ij}^2 - \bar{y}_i \sum_j y_{ij}$$

$$= \sum_i \sum_j (y_{ij} - \bar{y}_i)^2,$$

where, for convenience, we have used the particular solution with $\hat{\mu} = 0$. Next, we must assume H_0 is true, solve the normal equations and determine $Q_0 + Q_1$ as the sum of squares used to estimate σ^2 if H_0 is true. Given $\tau_1 = \tau_2 = \cdots = \tau_t = \tau$, say, the expected value of each observation is the same, and the two normal equations we have are identical and both equal

$$n(\hat{\mu} + \hat{\tau}) = \sum_i \sum_j y_{ij}$$

whose solution is $\hat{\mu} + \hat{\tau} = (1/n)\sum_i\sum_j y_{ij} = \bar{y}$, the overall mean; thus $\hat{\tau} = \bar{y} - \hat{\mu}$, where again $\hat{\mu}$ is arbitrary. Then, the sum of squares which would be used to estimate σ^2 (the *residual* sum of squares) is

$$Q_0 + Q_1 = \sum_i \sum_j y_{ij}^2 - \hat{\mu} \sum_i \sum_j y_{ij} - (\bar{y} - \hat{\mu}) \sum_i \sum_j y_{ij}$$

$$= \sum_i \sum_j (y_{ij} - \bar{y})^2$$

$$= \sum_i \sum_j (y_{ij} - \bar{y}_i)^2 + \sum_i n_i(\bar{y}_i - \bar{y})^2,$$

no matter what the value of $\hat{\mu}$. Thus we have $Q_1 = (Q_0 + Q_1) - Q_0 = \sum_i n_i(\bar{y}_i - \bar{y})^2$; the matrix of Q_0 is idempotent of rank $\sum_i (n_i - 1)$, and the matrix of Q_1 is idempotent of rank $k - 1$. Furthermore, the noncentrality parameter for Q_0 is 0, regardless of whether H_0 is true, and the noncentrality parameter of Q_1 is $\sum_i [n_i(\tau_i - \bar{\tau})^2/\sigma^2]$, and the products of the matrices of the two forms is $\mathbf{0}$. Thus, Q_0 is $\chi^2_{\Sigma(n_i-1)}$, Q_1 is χ'^2_{k-1} and the two are independent; the likelihood ratio test of H_0 then is equivalent to forming the ratio $[Q_1/(k-1)]/[Q_0/\sum(n_i - 1)]$ and rejecting H_0 if this test statistic exceeds $F_{k-1,\Sigma(n_i-1)}(1-\alpha)$. The test statistic has the $F'_{k-1,\Sigma(n_i-1)}$ distribution and, for any given set of values for τ_i such that H_0 is not true, the noncentral F distribution would be used to determine the power of the test. The test statistic has the central $F_{k-1,\Sigma(n_i-1)}$ distribution if and only if $\sum n_i(\tau_i - \bar{\tau})^2 = 0$, i.e., if and only if H_0 is true.

The quantities necessary to test that the treatments are equal in their effects are generally displayed in an analysis of variance as given in Table 7.4.1. The following example presents a numerical example of this type of design and the analysis of variance to test equality of treatment effect.

TABLE 7.4.1. Analysis of Variance for Testing H_0: $\tau_1 = \tau_2 = \cdots = \tau_k$ for Completely Randomized Design

Source	Degrees of freedom	Sums of squares	Mean squares
Mean (assuming equal treatments)	1	$n\bar{y}^2$	
Treatments (adjusted)	$k - 1$	$Q_1 = \sum_i n_i(\bar{y}_i - \bar{y})^2$	$Q_1/(k - 1)$
Residual	$\sum_i (n_i - 1)$	$Q_0 = \sum_i \sum_j (y_{ij} - \bar{y}_i)^2$	$Q_0/\sum_i (n_i - 1)$
Total	$\sum_i n_i$	$\sum_i \sum_j y_{ij}^2$	

Example 7.4.1

In designing a new piece of equipment, the relative location of the controls can frequently have a major effect on the ease of use or the safety of use of the equipment. To take an oversimplified example, suppose that a piece of equipment has only two control knobs; they could be mounted essentially anywhere on the instrument panel before the operator. Since there is freedom of choice of location, there are three major configurations to be considered: (1) mount both side by side on the right side of the panel; (2) mount both side by side on the left side of the panel; and (3) mount one on the right and one on the left side of the panel (in this case the two are separated). A sample is available of 17 people who are representative of the types of operators for which the machine is designed. Of these, six are assigned at random to each of configurations (1) and (2), and the remaining five are assigned to configuration (3). Each is then given a brief training session on the use of the equipment, with the assigned instrument configuration. Then, to compare the three configurations, each of the 17 is given the same sequence of operations to perform and the total elapsed performance time is recorded. The elapsed times required are presented in Table 7.4.2. Since the purpose is to compare the three configurations, the

TABLE 7.4.2. Elapsed Times Required (seconds)

Configuration	
(1)	627, 598, 612, 615, 601, 620
(2)	590, 605, 631, 610, 622, 603
(3)	597, 603, 601, 624, 617

TABLE 7.4.3. Analysis of Variance for Configurations

Source	Degrees of freedom	Sums of squares	Mean squares
Mean	1	6,333,022.1	
Configurations	2	39.0	19.5
Residual	14	2,204.9	157.5
Total	17	6,335,266	

resulting analysis of variance is presented in Table 7.4.3. With the assumption that the distribution of such times is normal, with the same variance σ^2 for each configuration, we would accept the hypothesis that the mean performance time is the same for all three configurations.

We shall conclude this section by examining a slightly more complex experimental design model: a two-way classification without interaction, a model which corresponds to a *randomized block experiment*. A simple example of this sort of experiment is as follows: Suppose we want to compare the yields of, say, four different types of hybrid corn and that the field available for making our comparison is known to have differing amounts of natural fertility going from the north end to the south end. The field then is cut into, say, five strips running from east to west, and each strip is further subdivided into four plots, one for each of the four different hybrids. By subdividing the field in this way it is reasonable to assume that the four plots within each strip, or block, are equally fertile, but that in going from one block to the next there is a difference in fertility. The standard randomized block model also assumes that the relative comparison of the four hybrids is unchanged from one block to the next; that is, if the true difference in yield between hybrids 1 and 2 is 10 bushels per acre in the southernmost block, it is also 10 bushels per acre in the northernmost block (as well as in the other 3 blocks in between). It may well be that the natural fertility causes the actual yields to be higher in the southern block than they are in the northern block, but it is assumed that the difference in yields is a constant from block to block. This is called no *interaction* between the types of corn (treatments) and the blocks. In actually performing the experiment, of course, the treatments are randomly allocated to the plots within each block.

To see how the comparison of treatments is effected in such a model, let us assume that we have t treatments to compare and b different blocks, each consisting of t experimental units. Within each block the treatments are randomly allocated to the experimental units. It is assumed that all the units in the same block are homogeneous but that units taken from different blocks have different expected values, even if the same treatment is applied.

If we let Y_{ij} represent the yield we will observe from the ith treatment in the jth block, then, we have

$$E[Y_{ij}] = \mu + \tau_i + \beta_j, \qquad i = 1, 2, \ldots, t; \qquad j = 1, 2, \ldots, b;$$

μ is again an overall mean parameter, τ_i is the amount added because the experimental unit received the ith treatment (again it may be positive or negative) and β_j is the amount added because the unit is located in the jth block. Note then that the expected difference in yield between treatments 1 and 2, in block 1, is $(\mu + \tau_1 + \beta_1) - (\mu + \tau_2 + \beta_1) = \tau_1 - \tau_2$; this expected difference in block 2 is $(\mu + \tau_1 + \beta_2) - (\mu + \tau_2 + \beta_2) = \tau_1 - \tau_2$. It is obvious that this expected difference is the same, no matter which block we look at; indeed, the expected value for any *comparison* of the treatment effects ($\sum_{i=1}^t c_i E[Y_{ij}] = \sum_{i=1}^t c_i \tau_i$, where $\sum_{i=1}^t c_i = 0$) is the same constant within any block. This is what is assumed when we say there is no interaction between blocks and treatments. It is also assumed that all of the Y_{ij}'s are independent, normal, and that they have the same variance σ^2.

To again put this model into our matrix notation, let $\mathbf{y}$ be the $bt \times 1$ vector whose first t components are $y_{11}, y_{21}, \ldots, y_{t1}$ (the observed yields from the first block), whose next t components are $y_{12}, y_{22}, \ldots, y_{t2}$ (yields from the second block), etc. The final t components of $\mathbf{y}$ are the observations from the bth block. Then, define $\mathbf{x}$ to be the $bt \times (b + t + 1)$ matrix whose first $t \times (b + t + 1)$ submatrix is

$$(\mathbf{1}, \mathbf{I}_t, \mathbf{1}, \mathbf{0}, \ldots, \mathbf{0})$$

where $\mathbf{1}$ is the $t \times 1$ vector each of whose components is 1, $\mathbf{0}$ is the $t \times 1$ null vector, and $\mathbf{I}_t$ is the $t \times t$ identity. The second $t \times (b + t + 1)$ submatrix in $\mathbf{x}$ is

$$(\mathbf{1}, \mathbf{I}_t, \mathbf{0}, \mathbf{1}, \mathbf{0}, \ldots, \mathbf{0}),$$

the third is $(\mathbf{1}, \mathbf{I}_t, \mathbf{0}, \mathbf{0}, \mathbf{1}, \ldots, \mathbf{0})$ etc. and the final one is

$$(\mathbf{1}, \mathbf{I}_t, \mathbf{0}, \mathbf{0}, \mathbf{0}, \ldots, \mathbf{0}, \mathbf{1}).$$

Then, defining $\boldsymbol{\beta}' = (\mu, \tau_1, \tau_2, \ldots, \tau_t, \beta_1, \beta_2, \ldots, \beta_b)$ we have $\mathbf{Y}$ is multivariate normal with mean $\mathbf{x}\boldsymbol{\beta}$ and covariance $\sigma^2\mathbf{I}$. Notice then that $\mathbf{x}$ has $b + t + 1$ columns, but if we sum the second through $(t + 1)$st columns together we get the first column (all 1's) and similarly if we sum the $(t + 2)$nd through $(b + t + 1)$st column we again get the first column. Thus

$$r(\mathbf{x}) \leq b + t - 1.$$

We can also by inspection see that the second through $(t + 1)$st columns are linearly independent within themselves (because of the placement of 1's and 0's) as are the $(t + 2)$nd through $(b + t + 1)$st columns; also, any two proper subsets of columns chosen from these two sets are independent and thus $r(\mathbf{x}) = b + t - 1$. To make this rank argument clear, notice that if $b = t = 2$, then the first 2×5 submatrix of $\mathbf{x}$ is $(\mathbf{1}, \mathbf{I}_2, \mathbf{1}, \mathbf{0})$ while the second is $(\mathbf{1}, \mathbf{I}_2, \mathbf{0}, \mathbf{1})$; thus we have

$$\mathbf{x} = \begin{bmatrix} 1 & 1 & 0 & 1 & 0 \\ 1 & 0 & 1 & 1 & 0 \\ \hdashline 1 & 1 & 0 & 0 & 1 \\ 1 & 0 & 1 & 0 & 1 \end{bmatrix}.$$

Clearly the second and third columns of $\mathbf{x}$ are linearly independent, as are the fourth and fifth. We also note that the sum of the second and third gives the first column as does the sum of the fourth and fifth. Thus, in the first three columns only two vectors are linearly independent; adding on the fourth and fifth columns only increases the rank by one, since their sum is already represented in the first three. Thus, this particular $\mathbf{x}$ matrix has rank $b + t - 1 = 2 + 2 - 1 = 3$; this argument clearly generalizes to any positive integer values for b and t. Thus, $\mathbf{x}'\mathbf{x}$, in the general case, has size $(b + t + 1) \times (b + t + 1)$ and is of rank $b + t - 1$, and again the normal equations have an infinite number of solutions (a twofold infinity, since the rank is two less than the size, meaning two functions of the parameters can be given arbitrary values and then the rest are specified).

Again, the normal equations to be solved for the maximum likelihood and least squares estimates are given by $\mathbf{x}'\mathbf{x}\hat{\mathscr{B}} = \mathbf{x}'\mathbf{y}$. We might here make use of a trick that is frequently useful in deriving the normal equations for design models. Note that $E[\mathbf{x}'\mathbf{Y}] = \mathbf{x}'\mathbf{x}\boldsymbol{\beta}$. Thus, if we can easily write out the elements of $\mathbf{x}'\mathbf{y}$, which give the right-hand sides of the normal equations, in scalar notation, then the corresponding left-hand sides will be given by their expected values, where a carat is placed over each parameter. Because of the simple pattern which occurs in most of the common design matrices, this method is easy to employ.

We recall that our $\mathbf{x}$ matrix for the randomized block design is

$$\mathbf{x} = \begin{bmatrix} \mathbf{1} & \mathbf{I}_t & \mathbf{1} & \mathbf{0} & \cdots & \mathbf{0} \\ \mathbf{1} & \mathbf{I}_t & \mathbf{0} & \mathbf{1} & \cdots & \mathbf{0} \\ \cdot & \cdot & \cdot & \cdot & & \cdot \\ \cdot & \cdot & \cdot & \cdot & & \cdot \\ \cdot & \cdot & \cdot & \cdot & & \cdot \\ \mathbf{1} & \mathbf{I}_t & \mathbf{0} & \mathbf{0} & \cdots & \mathbf{1} \end{bmatrix}$$

written in terms of submatrices. The first element of $\mathbf{x}'\mathbf{y}$ then is $\sum_i\sum_j y_{ij}$, since the first column of $\mathbf{x}$ has a 1 in each position; the second element is $\sum_j y_{1j}$ since the second column of $\mathbf{x}$ has a 1 in every row in which $i = 1$. Similarly, the third through $(t+1)$st elements of $\mathbf{x}'\mathbf{y}$ are respectively $\sum_j y_{2j}, \sum_j y_{3j}, \ldots, \sum_j y_{tj}$, since these columns only have a 1 in them when $i = 2$, $i = 3$, etc. Similarly, the $(t+2)$nd through $(b+t+1)$st elements of $\mathbf{x}'\mathbf{y}$ are, respectively, $\sum_i y_{i1}, \sum_i y_{i2}, \ldots, \sum_i y_{ib}$, since these columns only have 1's in them when $j = 1$, when $j = 2$, etc. Now, to get the left-hand sides for the normal equations we simply have to compute the expected values of these right-hand sides. Thus, the first left-hand side is

$$E\left[\sum_i \sum_j Y_{ij}\right] = \sum_i \sum_j (\mu + \tau_i + \beta_j) = bt\mu + b\sum_i \tau_i + t\sum_j \beta_j,$$

and the second is

$$E\left[\sum_j Y_{1j}\right] = \sum_j (\mu + \tau_1 + \beta_j) = b\mu + b\tau_1 + \sum_j \beta_j.$$

Clearly, the third through $(t+1)$st left-hand side is equal to $b\mu + b\tau_i + \sum_j \beta_j$, where i is set equal to $2, 3, \ldots, t$, in turn. Similarly, the next left-hand side is given by

$$E\left[\sum_i Y_{i1}\right] = \sum_i (\mu + \tau_i + \beta_1) = t\mu + \sum_i \tau_i + t\beta_1,$$

and the remaining ones are given by $t\mu + \sum_i \tau_i + t\beta_j$, where j equals $2, 3, \ldots, b$, in turn. Thus, our normal equations are

$$bt\hat{\mu} + b\sum_i \hat{\tau}_i + t\sum_j \hat{\beta}_j = \sum_i \sum_j y_{ij}$$

$$b\hat{\mu} + b\hat{\tau}_i + \sum_j \hat{\beta}_j = \sum_j y_{ij}, \qquad i = 1, 2, \ldots, t,$$

$$t\hat{\mu} + \sum_i \hat{\tau}_i + t\hat{\beta}_j = \sum_i y_{ij}, \qquad j = 1, 2, \ldots, b.$$

One set of solutions for these equations is

$$\hat{\mu} = \bar{y}_{..} = \frac{1}{bt}\sum_i \sum_j y_{ij},$$

$$\hat{\tau}_i = \bar{y}_{i.} - \bar{y}_{..} = \frac{1}{b}\sum_j y_{ij} - \bar{y}_{..}, \qquad i = 1, 2, \ldots, t,$$

$$\hat{\beta}_j = \bar{y}_{.j} - \bar{y}_{..} = \frac{1}{t}\sum_i y_{ij} - \bar{y}_{..}, \qquad j = 1, 2, \ldots, b.$$

(These solutions were arrived at by realizing that $\sum_i \tau_i$ and $\sum_j \beta_j$ are not estimable functions, since no linear combination of the rows of $\mathbf{x}$ will give either $(0, 1, 1, \ldots, 1, 0, 0, \ldots, 0)$ or $(0, 0, 0, \ldots, 0, 1, 1, \ldots, 1)$; thus their values are arbitrary and can be set equal to 0, for example. Setting them equal to 0 easily gives the solution mentioned.) The estimated difference between treatments k and m then is $\hat{\tau}_k - \hat{\tau}_m = \bar{y}_{k.} - \bar{y}_{m.}$. The unbiased estimate of σ^2 is

$$s^2 = \frac{Q_0}{bt - (b + t - 1)} = \frac{Q_0}{(b - 1)(t - 1)}$$

where

$$\begin{aligned} Q_0 &= \mathbf{y}'\mathbf{y} - \hat{\boldsymbol{\beta}}'\mathbf{x}'\mathbf{y} \\ &= \sum_i \sum_j y_{ij}^2 - \bar{y}_{..} \sum_i \sum_j y_{ij} - \sum_i \left[(\bar{y}_{i.} - \bar{y}_{..}) \sum_j y_{ij} \right] - \sum_j \left[(\bar{y}_{.j} - \bar{y}_{..}) \sum_i y_{ij} \right] \\ &= \sum_i \sum_j y_{ij}^2 - bt\bar{y}_{..}^2 - b \sum_i (\bar{y}_{i.} - \bar{y}_{..})^2 - t \sum_j (\bar{y}_{.j} - \bar{y}_{..})^2, \end{aligned}$$

since

$$\bar{y}_{..} = \frac{1}{bt} \sum_i \sum_j y_{ij}, \qquad \bar{y}_{i.} = \frac{1}{b} \sum_j y_{ij}, \qquad \bar{y}_{.j} = \frac{1}{t} \sum_i y_{ij}.$$

To test equality of treatment effects, H_0: $\tau_1 = \tau_2 = \cdots = \tau_t = \tau$, we use the same procedure as in the completely randomized design. We first must determine $Q_0 = \mathbf{y}'\mathbf{y} - \hat{\boldsymbol{\beta}}'\mathbf{x}'\mathbf{y}$, where $\hat{\boldsymbol{\beta}}$ is any estimate of the full vector of parameters with no restrictions; this quantity has already been given above in defining s^2. Next, we assume H_0 is true, and determine the sum of squares $Q_0 + Q_1$ which would be used to estimate σ^2. If H_0 is true

$$E[Y_{ij}] = \mu + \tau + \beta_j = \mu^* + \beta_j$$

no matter what the value of i. Note that this is exactly the model we used in the completely randomized design discussed earlier with $t = b$, $n_i = t$; thus the sum of squares used to estimate σ^2 would be

$$Q_0 + Q_1 = \sum_i \sum_j y_{ij}^2 - bt\bar{y}_{..}^2 - t \sum_j (\bar{y}_{.j} - \bar{y}_{..})^2 = \sum_i \sum_j (y_{ij} - \bar{y}_{.j})^2$$

as we had for that model. Then

$$Q_1 = (Q_0 + Q_1) - Q_0 = b \sum_i (\bar{y}_{i.} - \bar{y}_{..})^2.$$

By examining the matrices of Q_0 and Q_1 we can easily see that Q_0 is $\chi^2_{(b-1)(t-1)}$, Q_1 is χ'^2_{t-1} with noncentrality $b\sum_i (\tau_i - \bar{\tau})^2/\sigma^2$ and Q_0 and Q_1 are independent.

Thus, again $[Q_1/(t-1)]/[Q_0/(b-1)(t-1)]$ will have the $F'_{t-1,(b-1)(t-1)}$ distribution and the likelihood ratio test criterion is equivalent to rejecting H_0 if and only if $[Q_1/(t-1)]/[Q_0/(b-1)(t-1)] > F_{t-1,(b-1)(t-1)}(1-\alpha)$. The test statistic has the central F distribution if and only if H_0 is true. The quantities necessary to test the equality of treatment effects are generally displayed in an analysis of variance, like that given in Table 7.4.4.

TABLE 7.4.4. Analysis of Variance for Randomized Complete Block Design

Source	Degrees of freedom	Sums of squares	Mean squares
μ, β_j's if $\tau_1 = \tau_2 = \cdots = \tau_t$	b	$Q_2 = bt\bar{y}^2_{..} + t\sum(\bar{y}_{.j} - \bar{y}_{..})^2$	
τ_i's, adjusted	$t-1$	$Q_1 = b\sum_i(\bar{y}_{i.} - \bar{y}_{..})^2$	$Q_1/(t-1)$
Residual	$(b-1)(t-1)$	$Q_0 = \sum_i\sum_j y_{ij}^2 - Q_1 - Q_2$	$Q_0/(b-1)(t-1)$
Total	bt	$\sum_i\sum_j y^2_{ij}$	

EXERCISE 7.4

1. In the completely randomized design we found $Q_0 = \sum_i\sum_j(y_{ij} - \bar{y}_{i.})^2$. Write this as a quadratic form in the vector $\mathbf{y}$ of sample observations.

2. Verify the identity

$$\sum_i\sum_j(y_{ij} - \bar{y}_{..})^2 = \sum_i\sum_j(y_{ij} - \bar{y}_{i.})^2 + \sum_i n_i(\bar{y}_{i.} - \bar{y}_{..})^2.$$

3. Express all three quadratic forms given in question 2 in matrix notation.

4. Using the matrices of the quadratic forms in question 3, prove that Q_0 and Q_1 are independent, for a completely randomized design.

5. Verify the entries in Table 7.4.3.

6. In the randomized complete block design find the matrix for the quadratic form $Q_1 = b\sum_i(\bar{y}_{i.} - \bar{y}_{..})^2$.

7. For the randomized complete block design, find the matrix for the quadratic form

$$Q_0 = \sum_i\sum_j y^2_{ij} - bt\bar{y}^2_{..} - b\sum_i(\bar{y}_{i.} - \bar{y}_{..})^2 - t\sum_j(\bar{y}_{.j} - \bar{y}_{..})^2$$

and prove the independence of Q_0 and Q_1.

8. Assume that the operator of a large fleet of cars wants to use the brand of gasoline in his cars which gives the best mileage. Four major oil companies have stations in his area and each claims its regular gasoline gives the best mileage. His fleet of cars contains Chevrolets, Fords, and Plymouths. To compare the mileages of the four brands of gasoline, he chooses one car of each of the makes; each of these cars is driven over the same measured 100-mile course four times, once with a full tank of each brand of gasoline. The same driver is used for each of the 12 combinations. The following data (miles per gallon) resulted:

	Brand A	Brand B	Brand C	Brand D
Ford	16.58	17.21	17.02	16.85
Chevrolet	15.42	16.30	16.32	15.71
Plymouth	17.47	17.95	17.90	17.21

Do these data indicate the average miles per gallon are the same for the four brands?

9. It was mentioned that a randomized block design assumes *no interaction* between blocks and treatments. In many cases this is not a reasonable assumption and indeed a frequent purpose of experimentation is to investigate the possible existence of interaction. In Example 7.4.1 we discussed an experimental comparison of three different control configurations on a panel. In such a case it is quite possible that the difference in time necessary to perform the chosen tasks for, say, configurations 1 and 2, is not the same for left-handed and right-handed operators; we might then like to allow for *interaction* between the configurations and the right versus left-handedness of the operator. Let τ_i be the effect of the ith treatment (configuration), $i = 1, 2, 3$ and let ρ_j, $j = 1, 2$, represent the contribution from the operator (ρ_1 for right-handed, ρ_2 for left-handed people), and assume we have six right-handed and six left-handed operators available for the experiment; thus two right-handed people will be assigned to each of the three configurations as will two left-handed people. Then, to check whether there is interaction we would assume

$$Y_{ijk} = \mu + \tau_i + \rho_j + (\tau\rho)_{ij} + e_{ijk}, \qquad i = 1, 2, 3, \qquad j = 1, 2, \qquad k = 1, 2$$

where Y_{ijk} is the time required by the kth person, who is either right- or left-handed ($j = 1$ or 2) using the ith configuration. Note that the expected difference between configurations 1 and 2, for a right-handed person, is $\tau_1 - \tau_2 + (\tau\rho)_{11} - (\tau\rho)_{21}$ and for a left-handed person is $\tau_1 - \tau_2 + (\tau\rho)_{12} - (\tau\rho)_{22}$. If the $(\tau\rho)_{ij}$'s are not equal, these expected values are unequal. Thus the $(\tau\rho)_{ij}$'s are the interaction parameters and we would want to test

$$H_0\colon\ (\tau\rho)_{11} = \cdots = (\tau\rho)_{32}.$$

With the usual assumptions of normality and independence of the Y_{ijk}'s, derive the likelihood ratio test criterion for this hypothesis.

10. The randomized complete block design is an example of a *balanced* or *orthogonal design*. The numerical computations to test equality of treatment effects are quite simple in such cases, as long as all the observations planned are actually made.

If one observation is missing, say Y_{11}, the computations for the exact test become much more cumbersome, and the following approximate procedure is frequently employed. Assume y_{11} is missing and replace it by X; then the residual sum of squares, Q_0, is expressed as a function of X and the observed y_{ij}'s (this is of course a quadratic function of X). By differentiation, the value of X which minimizes Q_0 is found and then used as the observed value of Y_{11} and the usual analysis is performed, reducing the residual degrees of freedom by 1. Show that this procedure leads to

$$X = \frac{ty'_{1.} + by'_{.1} - y'_{..}}{(b - 1)(t - 1)}$$

where $y'_{1.}$ is the total of the available observations for treatment 1, $y'_{.1}$ is the total of the available observations from block 1, and $y'_{..}$ is the overall total of the available observations. Show that

$$E(X) = E[Y_{11}] = \mu + \tau_1 + \beta_1.$$

CHAPTER 8

Nonparametric Methods

8.1 Introduction

Many of the techniques we have studied assume that the underlying population sampled is normal in form, for exact results. It can be shown that in many cases these techniques are quite *robust* with respect to non-normality, especially for large samples. A technique is called *robust* if it leads to essentially correct conclusions and probability statements, even if the population sampled does not have the distribution assumed in deriving the technique. In the case of small samples, though, it is quite possible that these methods may not give very accurate approximations. Thus, there is some need for techniques of inference for non-normal populations. In this chapter we shall study some *nonparametric* or *distribution-free* methods; these procedures make a minimum of assumptions regarding the population distribution and are generally appropriate no matter what the particular form of that distribution.

8.2 Order Statistics and Procedures

Many nonparametric methods are based on the *order statistics* of the sample. Given a random sample $X_1, X_2, \ldots, X_n$, of a population random variable X, the order statistics are defined as follows: $X_{(1)}$, the first order statistic, is the smallest sample value, $X_{(2)}$ is the second smallest, etc., and $X_{(n)}$ is the largest sample value. $X_{(1)}, X_{(2)}, \ldots, X_{(n)}$ are called the order statistics of the sample. Note immediately then that for any sample outcome we must have $X_{(1)} \leq X_{(2)} \leq \cdots \leq X_{(n)}$ and the observed order statistics are given by the ranked sample values. Thus in any particular case $X_{(1)}$ may be equal to any one of the n sample random variables, $X_{(2)}$ may be equal to any of the remaining $n - 1$, etc. Especially when sampling from a continuous population, the joint distribution of the n order statistics takes a particularly simple form, which we shall now derive.

Let X be a continuous random variable and let $X_1, X_2, \ldots, X_n$ be a random sample of X. Then the joint density function for the sample random variables is $f_X(x_1)f_X(x_2) \cdots f_X(x_n)$. We shall now let $X_{(1)}$ equal the minimum of $X_1, X_2, \ldots, X_n$, $X_{(2)}$ equals the second smallest of $X_1, X_2, \ldots, X_n, \ldots$ up to $X_{(n)}$ equals the maximum sample value. Then, *for any particular sample outcome*, such as $X_{(1)} = X_3$, $X_{(2)} = X_1$, $X_{(3)} = X_4, \ldots, X_{(n)} = X_2$, the jacobian of the transformation will have the form

$$J = \begin{vmatrix} 0 & 0 & 1 & 0 & \cdots & 0 \\ 1 & 0 & 0 & 0 & \cdots & 0 \\ 0 & 0 & 0 & 1 & \cdots & 0 \\ \cdot & \cdot & \cdot & \cdot & & \cdot \\ \cdot & \cdot & \cdot & \cdot & & \cdot \\ \cdot & \cdot & \cdot & \cdot & & \cdot \\ 0 & 1 & 0 & 0 & \cdots & 0 \end{vmatrix};$$

that is, for any sample outcome there will be exactly one 1 in each row and exactly one 1 in each column, the particular placement of 1's being determined by the ordering in magnitude of $X_1, X_2, \ldots, X_n$. From the definition of a determinant, then, we will obviously have $J = 1$ or $J = -1$, and in either case $|J| = 1$, for each possible sample outcome. Since the transformation from $X_1, X_2, \ldots, X_n$ to $X_{(1)}, X_{(2)}, \ldots, X_{(n)}$ is many to one, we must compute the absolute value of the jacobian for each possibility ($|J| = 1$ for all of them), then make the substitution of variables in the joint density for each, and sum these joint densities times $|J|$ over all the possible sample outcomes that lead to $X_{(1)} = x_{(1)}$, $X_{(2)} = x_{(2)}, \ldots, X_{(n)} = x_{(n)}$, etc. Thus we have

$$\begin{aligned} f_{X_{(1)}, X_{(2)}, \ldots, X_{(n)}}(x_{(1)}, x_{(2)}, \ldots, x_{(n)}) &= \sum_{\substack{\text{Sample} \\ \text{orderings}}} f_X(x_{(3)})f_X(x_{(1)})f_X(x_{(6)}) \cdots f_X(x_{(2)})\,|J| \\ &= n!f_X(x_{(1)})f_X(x_{(2)}) \cdots f_X(x_{(n)}) \end{aligned}$$

where $x_{(1)} < x_{(2)} < x_{(3)} < \cdots < x_{(n)}$ and the joint density is 0, otherwise. There are exactly $n!$ different orderings which can occur for the sample outcomes and for each of these the substitution of variables in the density yields $f_X(x_{(1)})f_X(x_{(2)})\cdots f_X(x_{(n)})$, not necessarily in this order of the arguments, and since multiplication is commutative, the joint density has the form given above. Note that the arguments for the order statistics themselves are necessarily ordered in numerical magnitude, since it would be impossible for a sample outcome to yield $x_{(1)} \geq x_{(2)}$, for example. (Equality could occur with probability 0 because of the continuity of X.)

You are asked in the problems below to establish that the density of the ith order statistic is

$$f_{X_{(i)}}(x) = \frac{\Gamma(n+1)}{\Gamma(i)\Gamma(n-i+1)} f_X(x)[F_X(x)]^{i-1}[1-F_X(x)]^{n-i}, \qquad i = 1, 2, \ldots, n$$

for a sample of n from a continuous random variable with density $f_X(t)$ and distribution function $F_X(t)$. Note, then, that if we take a sample of n of a random variable X which is uniform on (0, 1), we have

$$f_X(t) = 1, \qquad 0 < t < 1$$

$$F_X(t) = t, \qquad 0 < t < 1$$

and thus

$$f_{X_{(i)}}(x) = \frac{\Gamma(n+1)}{\Gamma(i)\Gamma(n-i+1)} x^{i-1}(1-x)^{n-i}, \qquad 0 < x < 1.$$

That is, $X_{(i)}$ is a beta random variable with parameters $\alpha = i$, $\beta = n - i + 1$. We know immediately then that

$$E[X_{(i)}] = \frac{\alpha}{\alpha+\beta} = \frac{i}{n+1} \qquad i = 1, 2, \ldots, n;$$

since, for the uniform (0, 1) distribution, the value of $X_{(i)}$ is also the area under $f_X(x)$ (the population density) to the left of $X_{(i)}$, we find that the expected area under the density to the left of $X_{(i)}$ is $i/(n+1)$. The expected area under the density between the largest and smallest sample values (between $X_{(1)}$ and $X_{(n)}$) is $(n-1)/(n+1)$, for a sample of size n. These results can be generalized to essentially any continuous distribution by using the following theorem.

THEOREM 8.2.1. Assume X is a continuous random variable with density $f_X(t)$ and distribution function $F_X(t)$ which is monotonically increasing over the range of X. Then $Y = F_X(X)$ is a uniform random variable on (0, 1).

Proof: Since $0 \le F_X(t) \le 1$, for all real t, we can see immediately that

$$F_Y(t) = P(Y \le t) = 0 \qquad \text{for all } t \le 0$$

and

$$F_Y(t) = P(Y \le t) = 1 \qquad \text{for all } t \ge 1.$$

For

$$0 < t < 1, \qquad \begin{aligned} F_Y(t) &= P(F_X(X) \le t) \\ &= P(X \le F_X^{-1}(t)) \\ &= F_X(F_X^{-1}(t)) \\ &= t \end{aligned}$$

and the result is established.

You are asked in the exercises below to establish the following result.

THEOREM 8.2.2. Assume F_X is any monotonically increasing distribution function, over the range of the random variable X. Then, if Y is uniform on the interval (0, 1), $X = F_X^{-1}(Y)$ has distribution function $F_X(t)$.

Theorem 8.1.2 is called the *probability integral transform* and for many years was used to generate random samples of a continuous random variable X with distribution function $F_X(t)$. Note that if we can, by some means, generate a random sample of a uniform random variable Y on (0, 1), then $F_X^{-1}(Y_1), F_X^{-1}(Y_2), \ldots, F_X^{-1}(Y_n)$ will be a random sample of X. Other faster computational routines are commonly used today to generate random samples of a continuous random variable.

Example 8.2.1

In many applications the *percentiles* (see Exercise 1.3, problems 20–23) of a population are of interest and it is then desired to estimate them from a random sample. For a continuous population the 100pth *percentile*, ξ_p, is defined by $F_X(\xi_p) = p$, $0 \le p \le 1$. A frequent measure of the middle of a population is $\xi_{.5}$, the population *median*. Note that $\xi_{.5}$ has the property that 50% of the population values do not exceed it. $\xi_{.25}$, $\xi_{.5}$, and $\xi_{.75}$ are called the population *quartiles* and $\xi_{.1}, \xi_{.2}, \ldots, \xi_{.9}$ are the population *deciles*. The difference $\xi_{.75} - \xi_{.25}$, the *interquartile range*, is sometimes used as a measure of population variability instead of σ. The order statistics would seem natural choices to use in estimating particular population percentiles. For example, given a random sample from a continuous population, suppose we wanted to estimate the population median $\xi_{.5}$. If n is an odd number, then the middle sample value, $X_{((n+1)/2)}$, would seem a natural choice to estimate $\xi_{.5}$. The area under the population density to the left of $X_{((n+1)/2)}$ is $F_X(X_{((n+1)/2)})$. We shall compute $E[F_X(X_{((n+1)/2)})]$. If

$X_1, X_2, \ldots, X_n$ is a random sample of X, then $Y_i = F_X(X_i), i = 1, 2, \ldots, n$, is a random sample of a uniform (0, 1) random variable. Furthermore, since $F_X(t)$ is a nondecreasing function, $F_X(X_{(i)}) = Y_{(i)}$ are the order statistics for a uniform (0, 1) random variable. Then, from our results above, note that $E[F_X(X_{((n+1)/2)})] = E[Y_{((n+1)/2)}] = (n + 1)/2n = 1/2 + 1/2n$. Thus, the expected area to the left of the sample median, $X_{((n+1)/2)}$, for odd n, is not 1/2. Why doesn't this result imply that $E[X_{((n+1)/2)}] = \xi_{.5(1+(1/n))}$? See exercise 8 below.

Using the distributions for the order statistics of a random sample, many different types of inferences may be made. We shall study a few of them now. It was mentioned in Example 8.2.1 that percentiles are used in many applied areas. Let us now derive confidence intervals for the percentiles of a distribution, based on a random sample from the population. Recall that, if X is a continuous random variable and ξ_p is the 100pth percentile of its distribution, then $F_X(\xi_p) = p$. Now if $X_{(1)}, X_{(2)}, \ldots, X_{(n)}$ are the order statistics of a random sample of X, note that

$$\begin{aligned} P(X_{(k)} \leq \xi_p) &= P(F_X(X_{(k)}) \leq F_X(\xi_p)) \\ &= P(Y_{(k)} \leq p), \qquad k = 1, 2, \ldots, n \end{aligned}$$

where $Y_{(k)}$ is the kth order statistic of a uniform (0, 1) random variable. Thus

$$\begin{aligned} P(X_{(k)} \leq \xi_p) &= \int_0^p \frac{\Gamma(n+1)}{\Gamma(k)\Gamma(n-k+1)} x^{k-1}(1-x)^{n-k}\, dx \\ &= \sum_{i=k}^{n} \binom{n}{i} p^i (1-p)^{n-i}, \end{aligned}$$

from Exercise 1.3.14. This same result can also be established by using the reasoning in Exercise 8.2.4. The event $X_{(k)} \leq \xi_p$ will occur if and only if at least k sample values are less than or equal to ξ_p. Since $F_X(\xi_p) = p$, and our sample values are independent, our random sample of n values of X is equivalent to n independent Bernoulli trials with probability p of success on each trial. Thus, if we let W be the number of sample values that do not exceed ξ_p, W is a binomial random variable with parameters n and p and

$$P(X_{(k)} \leq \xi_p) = P(W \geq k) = \sum_{i=k}^{n} \binom{n}{i} p^i (1-p)^{n-i}$$

as we just derived above.

This probability statement for $X_{(k)}$ could, of course, be used to make a one-sided confidence interval for ξ_p; note that, for $k < m$, the events $\{X_{(k)} \leq \xi_p < X_{(m)}\}$ and $\{X_{(k)} \leq X_{(m)} \leq \xi_p\} = \{X_{(m)} \leq \xi_p\}$ are mutually exclusive and their union is $\{X_{(k)} \le \xi_p\}$. Thus

$$\begin{aligned} P(X_{(k)} \le \xi_p < X_{(m)}) &= P(X_{(k)} \le \xi_p) - P(X_{(m)} \le \xi_p) \\ &= \sum_{i=k}^{n} \binom{n}{i} p^i(1-p)^{n-i} - \sum_{i=m}^{n} \binom{n}{i} p^i(1-p)^{n-i} \\ &= \sum_{i=k}^{m-1} \binom{n}{i} p^i(1-p)^{n-i}. \end{aligned}$$

This statement can be used to compute two-sided confidence intervals for ξ_p.

Example 8.2.2

Assume that a random sample of $n = 10$ high school seniors take a standard quantitative reasoning exam that is scored on a "continuous" scale from 0 to 800. Then if $X_{(2)}$ and $X_{(9)}$ are the second and ninth order statistics of the 10 scores, the probability that they cover the population median, $\xi_{.5}$, is

$$\begin{aligned} P(X_{(2)} \le \xi_{.5} < X_{(9)}) &= \sum_{i=2}^{8} \binom{10}{i} \left(\frac{1}{2}\right)^i \left(1 - \frac{1}{2}\right)^{10-i} \\ &= .9786. \end{aligned}$$

Thus, if we observe $x_{(2)} = 427$ and $x_{(9)} = 692$, for example, we are 98% sure that the interval (427, 692) will cover the median $\xi_{.5}$ of the population test scores.

This two-sided confidence interval, or the one-sided confidence interval, could also be used to test hypotheses about the population percentiles. Suppose we wanted to test H_0: $\xi_{.4} = \xi^0$ versus H_0: $\xi_{.4} \ne \xi^0$, based on a random sample of n observations from the population. Then, by examining tables of the binomial distribution, assume we can find k and m, $k < m$, such that

$$\sum_{i=k}^{m-1} \binom{n}{i} (.4)^i(.6)^{n-i} = 1 - \alpha,$$

i.e.,

$$P(X_{(k)} \le \xi_{.4} < X_{(m)}) = 1 - \alpha.$$

We then observe our m sample values and if $\xi^0 < X_{(k)}$ or $\xi^0 \ge X_{(m)}$ we reject H_0. We clearly have a probability of type I error equal to α. Note that n, k and m must all be specified in advance. To make a one-sided test we use the same procedure with a one-sided confidence interval for the percentile of interest.

In many applications, it is of interest to derive a random interval, based on a random sample of n observations, which has a known probability $1 - \alpha$ of including at least the proportion p of the population values. Such an interval is called a *tolerance interval.* If $X_{(i)}$ and $X_{(j)}$ are two order statistics from the sample, $i < j$, the random proportion of the population which is included between them is $W_{ij} = F_X(X_{(j)}) - F_X(X_{(i)}) = Y_{(j)} - Y_{(i)}$, where $Y_{(j)}$ and $Y_{(i)}$ are the order statistics of a sample of a uniform (0, 1) random variable. Thus, if we can find the distribution for W_{ij}, it could be used to find tolerance intervals for the distribution of X.

To derive the distribution of W_{ij}, for any choice of i and j, let us first derive the joint distribution of what are called the *coverage random variables.* Thus, we assume we have a random sample of n observations of a continuous random variable X and define

$$V_1 = F_X(X_{(1)}) = Y_{(1)}$$

$$V_2 = F_X(X_{(2)}) - F_X(X_{(1)}) = Y_{(2)} - Y_{(1)}$$

$$\cdots\cdots\cdots\cdots\cdots\cdots\cdots\cdots\cdots$$

$$V_n = F_X(X_{(n)}) - F_X(X_{(n-1)}) = Y_{(n)} - Y_{(n-1)}$$

where again $Y_{(1)}, Y_{(2)}, \ldots, Y_{(n)}$ are the uniform order statistics. Note that V_i, $i = 2, 3, \ldots, n$ provides the amount of the distribution of X that is "covered" by two successive order statistics; thus the V_i's are called *coverage* random variables.

The joint density of $Y_{(1)}, Y_{(2)}, \ldots, Y_{(n)}$ is

$$f_{Y_{(1)},Y_{(2)},\ldots,Y_{(n)}}(y_1, y_2, \ldots, y_n) = n! \qquad 0 < y_1 < y_2 < \cdots < y_n < 1.$$

If we define

$$v_1 = y_1$$

$$v_2 = y_2 - y_1$$

$$\cdots\cdots\cdots\cdots$$

$$v_n = y_n - y_{n-1}$$

note that the inverse transformation is

$$y_1 = v_1$$

$$y_2 = v_1 + v_2$$

$$y_3 = v_1 + v_2 + v_3$$

$$\cdots\cdots\cdots\cdots\cdots\cdots$$

$$y_n = v_1 + v_2 + \cdots + v_n$$

and thus the jacobian of the transformation is 1. Then the joint density of $V_1, V_2, \ldots, V_n$ is

$$f_{V_1,V_2,\ldots,V_n}(v_1, v_2, \ldots, v_n) = n!$$

$$\text{for } v_i > 0, \qquad i = 1, 2, \ldots, n, \qquad \sum_{i=1}^{n} v_i < 1.$$

Note that this density is symmetric in $v_1, v_2, \ldots, v_n$; thus the density function of the sum of any particular $r < n$ of the coverages $V_1, V_2, \ldots, V_n$ is the same as the density of the sum of any other set of r of them. For example, if $j = i + r$, the density of $V_{i+1} + V_{i+2} + \cdots + V_j = Y_{(j)} - Y_{(i)}$ must be the same as the density of $V_1 + V_2 + \cdots + V_r = Y_{(j-i)}$, the $(j - i)$th order statistic from a uniform (0, 1) distribution. Thus the density for

$$W_{ij} = F_X(X_{(j)}) - F_X(X_{(i)})$$

is completely specified by the difference $j - i = r$ and this density is the same as that of $Y_{(j-i)}$, namely

$$f_{Y_{(j-i)}}(y) = \frac{\Gamma(n+1)}{\Gamma(j-i)\Gamma(n-j+i+1)}\, y^{j-i-1}(1-y)^{n-j+i}, \qquad 0 < y < 1,$$

a beta density with parameters $\alpha = j - i$ and $\beta = n - j + i + 1$. The expected coverage, then, between the ith and jth order statistics is

$$(j - 1)/(n + 1).$$

The following example illustrates the computation of a particular tolerance interval.

Example 8.2.3

An industrial organization has just received a new piece of equipment for manufacturing machined parts. The parts to be made by the machine have one critical dimension. It is recognized that the value of this dimension will naturally vary from one manufactured part to the next, but it is important that the great majority of the parts which will be made should have this dimension within a specified interval. Note that if n parts are manufactured on the equipment and their dimensions measured, the probability that the interval $(X_{(1)}, X_{(n)})$ includes at least 80% of the distribution of dimensions is

$$P(Y_{(n-1)} \geq .80) = \int_{.80}^{1} \frac{\Gamma(n+1)}{\Gamma(n-1)\Gamma(2)}\, y^{n-2}(1-y)\, dy$$
$$= (n-1)(.8)^{n-1} + 1 - .8^n.$$

This equation can be used to determine the n which will make this probability as large as is desired. The corresponding number of parts could then be manufactured, and $x_{(1)}$ and $x_{(n)}$ determined to see if the equipment may be expected to perform satisfactorily.

EXERCISE 8.2

1. Suppose we have a discrete population with distribution $P(X = x) = 1/3$, $x = 1, 2, 3$. Let X_1, X_2 be a random sample of X and derive the joint distribution and marginal distributions of

$$X_{(1)} = \min(X_1, X_2), \qquad X_{(2)} = \max(X_1, X_2).$$

2. Assume a finite population with elements 1, 2, 3 and, for a sample of 2 without replacement, derive the joint distribution and marginal distributions for $X_{(1)} =$ minimum sample value, $X_{(2)} =$ maximum sample value.

3. By integrating the joint density function for the order statistics for a sample of size n of a continuous random variable X, show that the density of the ith order statistic is

$$f_{X_{(i)}}(x) = \frac{\Gamma(n+1)}{\Gamma(i)\Gamma(n-i+1)} f_X(x)[F_X(x)]^{i-1}[1 - F_X(x)]^{n-i}, \qquad i = 1, 2, \ldots, n.$$

(Hint: The region of integration is $-\infty < x_{(1)} < x_{(2)} < \cdots < x_{(i-1)} < x_{(i)}$ and $x_{(i)} < x_{(i+1)} < \cdots < x_{(n)} < \infty$.)

4. The following direct method of argument may be used to establish the distribution function, and then the density function, for $X_{(i)}$ defined in question 3 above. Note that $F_{X_{(i)}}(t) = P(X_{(i)} \le t) = P$(at least i sample values are less than or equal to t); the probability any given sample value does not exceed t is $F_X(t)$. Establish $f_{X_{(i)}}(t)$ in this way.

5. By integrating the joint density function for the order statistics for a random sample of size n of a continuous random variable X, show that the joint density function for $X_{(i)}$ and $X_{(j)}$, $i < j$ is

$$\begin{aligned} f_{X_{(i)},X_{(j)}}(x_{(i)}, x_{(j)}) = {} & \frac{\Gamma(n+1)}{\Gamma(i)\Gamma(j-1)\Gamma(n-j+1)} f_X(x_{(i)}) f_X(x_{(j)}) \\ & \times [F_X(x_{(i)})]^{i-1}[F_X(x_{(j)}) - F_X(x_{(i)})]^{j-i-1} \\ & \times [1 - F_X(x_{(j)})]^{n-j}, \end{aligned}$$

where $-\infty < x_{(i)} < x_{(j)} < \infty$. (Hint: The region of integration is $-\infty < x_{(1)} < \cdots < x_{(i)}$, $x_{(i)} < x_{(i+1)} < \cdots < x_{(j)}$, and $x_{(j)} < x_{(j+1)} < \cdots < x_{(n)} < \infty$.)

6. Describe how the joint distribution function, and hence the joint density function, for $X_{(i)}$ and $X_{(j)}$, might be derived by a direct argument using multinomial trials, analogous to question 4.

7. Given a random sample of size n of a continuous random variable X, where n is even, the *median* of the sample is

$$M = \frac{1}{2}[X_{(n/2)} + X_{(n/2+1)}].$$

Find bounds for the expected area under the population density to the left of M.

8. Given X is exponential with parameter $\lambda = 1$, find the density for $X_{(2)}$ from a sample of size 3 and use it to compute $E[X_{(2)}]$. Compare $F_X(E[X_{(2)}])$ and $E[F_X(X_{(2)})]$.

9. How large a sample, n, should we select of a continuous random variable X so that the expected area under the density of X between $X_{(1)}$ and $X_{(n)}$ is .99?

10. The *sample range* is $R = X_{(n)} - X_{(1)}$. If X is uniform (0, 1), show that the density of R is

$$f_R(r) = n(n-1)(1-r)r^{n-2} \qquad 0 < r < 1$$

and, for the value of n determined in 9, find the probability that $R = X_{(n)} - X_{(1)}$ includes at least .99 of the density (i.e., evaluate $P(R \geq .99)$).

11. If Z is a standard normal random variable and Z_1, Z_2 is a random sample of Z, show that $E[Z_{(2)}] = 1/\sqrt{\pi}$. What does this expected value become if Z has arbitrary mean μ and variance σ^2?

12. Assume X is uniform on $[\theta - (1/2), \theta + (1/2)]$ and $X_1, X_2, \ldots, X_n$ (n odd) is a random sample of X. Find the density of the sample median $X_{((n+1)/2)}$.

13. Let $X_1, X_2, \ldots, X_n$ be as defined in question 12 and find the density for

$$V = \tfrac{1}{2}(X_{(1)} + X_{(n)}).$$

Would you prefer $X_{((n+1)/2)}$ or V as an estimator for θ?

8.3 Permutations and Procedures

Let us now derive another basic distribution which can be used to make several types of nonparametric tests. Suppose we have n_1 symbols of one type, say a, and n_2 symbols of a second type, say b, arranged in a row; it is convenient to label the symbols so that $n_1 \leq n_2$ and to let $n = n_1 + n_2$ be the total number of symbols used. Clearly, if these n symbols are laid in a row at random, there are $\binom{n}{n_1}$ different sequences which could occur, since any n_1 of the n positions could be chosen for the a's and then the rest of the positions are filled with the b's. If these symbols are laid down randomly, then every possible one of these distinct sequences has the same probability, $1/\binom{n}{n_1}$, of occurring. This basic *permutation* distribution has many applications.

Within any one of these sequences, *runs* of the two types of symbols will occur, where a *run* is defined to be a block of consecutive identical symbols. Thus, in *aababbb* we have $n_1 = 3$, $n_2 = 4$, $n = 7$ and there are two runs of a's and two runs of b's. It is obvious, letting R_a be the number of runs of a's and R_b the number of runs of b's, that $R_a = R_b - 1$, R_b, or $R_b + 1$, since any two runs of a's must be separated by a run of b's, and vice versa. We shall first derive the joint probability function for R_a and R_b, and then use it to express the probability function for $R = R_a + R_b$, the total number of runs in such a sequence.

First, let us compute $P(R_a = R_b = i)$; to do this we need to count the number of sequences that can occur with the same number i of runs of the two types of symbols and then divide this quantity by $\binom{n}{n_1}$. We want to count all sequences which contain i runs of a's and of b's. First, we realize that each such sequence may start with either a run of a's or a run of b's and that there are equal numbers of such sequences. Thus, we need only count those sequences that start with a run of a's and multiply this number by 2. Clearly, if we have placed the n_1 a's in a row, there are $n_1 - 1$ interior positions between adjacent pairs of a's, into which $i - 1$ blocks of b's could be inserted (one block of b's must be placed at the end of the sequence) and each of these has resulted in splitting the a's into exactly i blocks; thus the number of ways in which the a's could be divided into exactly i blocks is $\binom{n_1 - 1}{i - 1}$. By exactly the same reasoning, $i - 1$ of the blocks of a's could be inserted between any of the $n_2 - 1$ adjacent pairs of b's to create the blocks of b's, so the b's can be split into exactly i blocks in $\binom{n_2 - 1}{i - 1}$ ways. The total number of sequences then with exactly i runs of each of the two types of symbols is

$$2\binom{n_1 - 1}{i - 1}\binom{n_2 - 1}{i - 1}$$

and thus

$$P(R_a = R_b = i) = \frac{2\binom{n_1 - 1}{i - 1}\binom{n_2 - 1}{i - 1}}{\binom{n}{n_1}}, \qquad i = 1, 2, \ldots, n_1.$$

Now let us count the number of sequences with i runs of a's and $i + 1$ runs of b's. Every such sequence must both begin and end with a run of b's. Counting in exactly the same way as above, the a's can be split into exactly i blocks in $\binom{n_1 - 1}{i - 1}$ ways and the b's can be split into exactly $i + 1$ blocks in

$\binom{n_2 - 1}{i}$ ways. Thus

$$P(R_a = R_b - 1 = i) = \frac{\binom{n_1 - 1}{i - 1}\binom{n_2 - 1}{i}}{\binom{n}{n_1}}, \qquad i = 1, 2, \ldots, n_1.$$

It remains for us to count the number of sequences with $i + 1$ runs of a's and i runs of b's; each of these sequences must both begin and end with a run of a's. The a's can be separated into exactly $i + 1$ blocks in $\binom{n_1 - 1}{i}$ ways and the b's can be separated into exactly i blocks in $\binom{n_2 - 1}{i - 1}$ ways. Thus

$$P(R_a = R_b + 1 = i + 1) = \frac{\binom{n_1 - 1}{i}\binom{n_2 - 1}{i - 1}}{\binom{n}{n_1}}, \qquad i = 1, 2, \ldots, n_1.$$

Using these results it is quite straightforward to write down the distribution of $R = R_a + R_b$, the total number of runs in such sequences. In fact

$$P(R = 2i) = \frac{2\binom{n_1 - 1}{i - 1}\binom{n_2 - 1}{i - 1}}{\binom{n}{n_1}} \qquad i = 1, 2, \ldots, n_1$$

$$P(R = 2i + 1) = \frac{\binom{n_1 - 1}{i - 1}\binom{n_2 - 1}{i} + \binom{n_1 - 1}{i}\binom{n_2 - 1}{i - 1}}{\binom{n}{n_1}},$$

$$i = 1, 2, \ldots, n_1.$$

This distribution can be used for many types of nonparametric tests, two examples of which are given below. For small values of n_1 and n_2 exact tables of this distribution are available and, if $n_1 \geq 10$, the distribution is well approximated by a normal density with

$$\mu = \frac{2n_1n_2}{n}$$

$$\sigma^2 = 4\,\frac{n_1^2n_2^2}{n^3}.$$

Example 8.3.1

The following test of randomness of the elements of a sample is sometimes used. Assume that $X_1, X_2, \ldots, X_n$ is a random sample of a random variable and, preserving the order in which the observations were obtained, let

$$\begin{aligned} Y_i &= a \qquad \text{if } X_i > M, \text{ the sample median} \\ &= b \qquad \text{if } X_i \leq M \end{aligned}$$

for $i = 1, 2, \ldots, n$. Thus we have replaced the sample elements by a sequence of n_1 a's and n_2 b's, where $n_1 = n_2 = n/2$, if n is even and $n_1 = (n-1)/2$, $n_2 = (n+1)/2$ if n is odd. If the sample is truly random, the distribution of the total number of runs is the one just derived and, for a random sample we would not expect to observe either too few or too many runs. If we have very few runs in the sequence, there may be a trend in the sample values (most of the early observations are below the median and the later ones above it, for example, or vice versa) and thus we might not be willing to accept $X_1, X_2, \ldots, X_n$ as a truly random sample. Similarly, if we observe too many runs it may be indicative of a cyclic effect, that consecutive observations are above and below M. Thus, the critical region generally employed for this test of randomness is to reject randomness of the sample values if either $R \leq r_{\alpha/2}$ or $R \geq r_{1-(\alpha/2)}$, where R is the total number of runs observed. The critical points $r_{\alpha/2}$ and $r_{1-(\alpha/2)}$ are either determined from tables of the exact distribution given above or from the normal approximation to this distribution if n is sufficiently large.

Example 8.3.2

The distribution for the number of runs can also be used to test that two random samples were selected from the same population. Suppose $X_1, X_2, \ldots, X_{n_1}$ is a random sample of a random variable X and $Y_1, Y_2, \ldots, Y_{n_2}$ $(n_1 \leq n_2)$ is a random sample of a random variable Y. If in fact the distribution functions for X and Y are equal, we should have a random sample of X's (a's) and Y's (b's) if we combine the two samples and rank them in order of magnitude. If, on the other hand, the two distributions are not the same, we would expect to observe a fairly small number of runs in the ranked, combined sample, because then the X and Y values would not tend to be well mixed. Thus to use R as the test statistic, we would reject the hypothesis that the two distributions are equal if $R \leq r_\alpha$, where again r_α is determined from either tables of the exact distribution or from the normal approximation to the exact distribution.

In addition to the distributions of runs and tests based on them, the basic distribution for the permutations of different symbols can be used in many other ways. One of these is the *randomization test* for comparing two (or more) populations. Suppose we have a random sample $X_1, X_2, \ldots, X_{n_1}$ of a random variable X and a random sample $Y_1, Y_2, \ldots, Y_{n_2}$, $n_1 \leq n_2$, of a second random variable Y and we want to test the hypothesis that X and Y have the same distribution. Let $\bar{X} = (1/n_1)\sum X_i$ be the mean of the X values as observed and let $\bar{Y} = (1/n_2)\sum Y_i$ be the mean of the Y values and define

$$\tilde{D} = |\bar{X} - \bar{Y}|.$$

Now, if in fact the two distributions are the same, any one of the observed X values could equally as well have been observed as a Y value and vice versa. Thus we have available $n = n_1 + n_2$ observed values, any n_1 of which could have been X values, and the remainder Y values, if the two distributions are the same. For each possible selection of n_1 values from the n, we can compute their average $\bar{X}$ and the average of the remainder $\bar{Y}$ and again compute

$$D = |\bar{X} - \bar{Y}|$$

for each of these possible selections. Thus, we would have in total $\binom{n}{n_1}$ different values of D, possibly not all distinct, which are equally likely to occur under the assumption that the two distributions are identical, from which we can compute the distribution of D, given the observed values. If $\tilde{D}$, computed from the X's and Y's as observed, falls below $d_{\alpha/2}$ or above $d_{1-(\alpha/2)}$, the $100(\alpha/2)$ and $100(1 - (\alpha/2))$ percentiles of the distribution of D, we reject H_0. The following simple example illustrates this test.

Example 8.3.3

As an unrealistically simple example of the details of the use of this test, consider the following: John drives his own car to work on every work day. Thus there would be a theoretical distribution describing the time necessary for him to make this trip on Mondays, say, (distribution of X) and a conceivably different distribution of times it takes him to make this same trip on Thursdays (distribution of Y). Assume that for three successive weeks he observes $x_1 = 1233$, $x_2 = 1149$, and $x_3 = 1320$ (in seconds) and $y_1 = 1143$, $y_2 = 1338$, $y_3 = 1455$ as his driving times for Mondays and Thursdays. Then, assuming that these values are random samples from the distributions of X and Y, respectively, the observed $\bar{x} = 1234$ and $\bar{y} = 1312$ so $\tilde{d} = |\bar{x} - \bar{y}| = 78$. If in fact the two distributions are identical, then any 3 of the 6 observed numbers could have been an X observation. Note that in a case such as this where $n_1 = n_2$ $(= 3)$, we really only have to table $(1/2)\binom{6}{3} = 10$ of the possible sets of 3 values that could have been X values, since the remaining 3 are then assigned to be Y values and the remaining 10 sets of 3 will give the same values for $d = |\bar{x} - \bar{y}|$ (except $\bar{x}$ and $\bar{y}$ are interchanged). Table 8.3.1 gives the 10 subsets of the 6 numbers which include 1143 as an X value, the resulting $\bar{x}$, $\bar{y}$ and d. We can see then that, over all possible

TABLE 8.3.1. Possible x Samples

Values of x	$\bar{x}$	$\bar{y}$	d
1143, 1149, 1233	1175	1371	196
1143, 1149, 1320	1204	1342	138
1143, 1149, 1338	1210	1336	126
1143, 1149, 1455	1249	1297	48
1143, 1233, 1320	1232	1314	82
1143, 1233, 1338	1238	1308	70
1143, 1233, 1455	1277	1269	8
1143, 1320, 1338	1267	1279	12
1143, 1320, 1455	1306	1240	66
1143, 1338, 1455	1312	1234	78

partitions into 3 x and 3 y values, for this sample, d equals 8, 12, 48, 66, 70, 78, 82, 126, 138, and 196, each with probability 1/10; since our observed absolute difference $\tilde{d} = 78$ falls close to the median of this distribution, we would not reject the hypothesis based on this set of sample outcomes.

It is obvious that the amount of computation necessary to carry out this test becomes enormous if n_1 and n_2 are very large. However, with modern computer equipment and well-thought-out algorithms, it is within the realm of practicality today. One important reason the test is of interest is that Lehman and Stein (*Annals of Mathematical Statistics*, 1949) have shown that this randomization test is as powerful as the t test for equality of means (assuming equal variances) when sampling from normal populations. When sampling from non-normal populations, it may very well be more powerful.

Let us close this section by examining a frequently used nonparametric test that two samples come from the same distribution. This test was first described by Wilcoxon in the case of equal size samples from the two populations and later was extended to unequal sample sizes by Mann and Whitney; they also established the approximate normality of the test statistic for moderate sample sizes. Thus the test is commonly referred to as the *Wilcoxon-Mann-Whitney test* or more simply the *rank-sum test*.

Assume we have a random sample of size n_1 of X and a random sample of size n_2 of Y (both X and Y are continuous); again we desire to test the equality of the distributions for X and for Y. We combine the two samples and rank the $n = n_1 + n_2$ numbers in order of magnitude and then replace the actual observations by their ranks, $1, 2, \ldots, n$. Wilcoxon then suggested using the statistic T which is the sum of the ranks of the Y observations to test the equality of the two distributions.

Mann and Whitney proposed the following statistic, which we shall see is a linear function of T. Instead of ranking the combined sample values, form all possible pairs (X_i, Y_j), $i = 1, 2, \ldots, n_1$, $j = 1, 2, \ldots, n_2$ and define

$$Z_{ij} = 1 \qquad \text{if } X_i < Y_j$$

$$Z_{ij} = 0 \qquad \text{if } X_i > Y_j$$

and then let

$$U = \sum_i \sum_j Z_{ij};$$

Mann and Whitney proposed using U to test the equality of the two distributions. Note that $\sum_i Z_{ij}$, for fixed j, gives the total number of X observations which are smaller than Y_j, that is, the number of times Y_j is preceded by an X value in the combined ranked sample. Then, in adding over j as well we simply have accumulated these totals for all the Y values, giving the total number of times a Y is preceded by an X in the combined sample.

To see the relationship between T and U, let R_1 be the rank of the smallest Y value in the combined sample, R_2 the rank of the second smallest, etc. Then the smallest Y value is preceded by $R_1 - 1$ X values, the second smallest is preceded by $R_2 - 2$ X values, up to the largest Y value is preceded by $R_{n_2} - n_2$ X values. Thus, we have

$$U = \sum_{i=1}^{n_2} (R_i - i) = \sum_{i=1}^{n_2} R_i - \sum_{i=1}^{n_2} i = T - \frac{n_2(n_2 + 1)}{2}.$$

Thus the two statistics, U and T, are linearly related and any procedure based on one can easily be translated into a procedure based on the other.

The following difference equation has been used to table the exact distribution for U, for $n_1 \leq n_2 \leq 8$. Consider any fixed values of n_1, n_2 and $U = u$. Let $N(u; n_1, n_2)$ represent the number of different possible samples that give rise to $U = u$. Then, such a combined sample could have either an X value or a Y value as its largest member; if the largest value is an X, the value of U is unchanged by deleting that largest value, whereas, if the largest sample value is a Y, then the value of U is decreased by n_1 if that largest sample value is deleted. Thus $N(u; n_1, n_2) = N(u; n_1 - 1, n_2) + N(u - n_1; n_1, n_2 - 1)$; if the two distributions are identical then all possible $\binom{n}{n_1}$ sample arrangements are equally likely and we have

$$P(U = u) = N(u; n_1, n_2) \Big/ \binom{n}{n_1}$$

which can then be used to construct the distribution for U, assuming H_0 is true. If $n_1 = n_2 = 1$, then U can equal only 0 or 1 and $N(0; 1, 1) = N(1; 1, 1) = 1$. By using these initial conditions, the above difference equation can then be used to build up the distribution of U for any desired values of n_1 and n_2. To test H_0, of course we would find $u_{\alpha/2}$ and $u_{1-(\alpha/2)}$ and reject H_0 if $U < u_{\alpha/2}$ or $U > u_{1-(\alpha/2)}$, to have probability of type I error equal to α (at most). For $n_1 \leq n_2 \leq 8$ this exact distribution has been tabled. Remarkably, if both n_1 and n_2 are bigger than 8, the distribution of

$$Z = \frac{U + \dfrac{1}{2} - \dfrac{n_1 n_2}{2}}{\sqrt{\dfrac{1}{12} n_1 n_2 (n_1 + n_2 + 1)}}$$

is well approximated by the standard normal. This fact can be used to find good approximations to the critical values, $u_{\alpha/2}$ and $u_{1-(\alpha/2)}$.

Example 8.3.4

Let us use the data presented in Example 8.3.3 to illustrate the computation of the observed value of U and to use the Mann-Whitney test of the equality of the two distributions. Note that $\sum_i z_{i1} = 0$, $\sum_i z_{i2} = 3$, $\sum_i z_{i3} = 3$ and thus the observed value of U is $u = 6$. With $n_1 = n_2 = 3$ we find from tables of the distribution of U that $u_{.05} = 1$, $u_{.95} = 8$; that is, we should reject equality of the distribution of X and Y if we observe $u < 1$ or $u > 8$ for $\alpha = .10$. Thus we accept H_0 with the given sample values, using this test.

EXERCISE 8.3

1. Use the data given in Exercise 7.4.8 to compare mileage with Brand A and Brand B, using the randomization test. Also apply the Wilcoxon-Mann-Whitney test.

2. Five lightbulbs of type A were found to have lifetimes 550, 490, 525, 530, and 507 hours, and four lightbulbs of type B had lifetimes 580, 510, 520, and 501 hours. Use the randomization test to test that these two types of bulbs have the same lifetime distribution.

3. The exchange rate between the United States dollar and the Brazilian cruzeiro is periodically adjusted. Between August 27, 1968 and March 19, 1971, the rate was adjusted 21 times. The number of days between changes were 27, 55, 57, 57, 43, 55, 55, 51, 37, 41, 34, 40, 56, 49, 52, 14, 56, 48, 33, 49, 38. By counting runs above and below the median, would you conclude that these values may be looked at as the observed values of a random sample of a random variable?

4. During the years 1926 to 1945, inclusive, the number of deaths in the United States attributed to tornadoes were 144, 540, 92, 274, 179, 36, 394, 362, 47, 70, 552, 29, 183, 87, 65, 53, 384, 58, 275, 210; from 1946 to 1968, inclusive, the number of tornado deaths were 78, 313, 140, 212, 71, 34, 230, 516, 35, 125, 83, 191, 66, 57, 48, 51, 28, 31, 73, 299, 105, 116, 131. If we assume these two sequences are random samples of sizes $n_1 = 20$, $n_2 = 23$, of random variables X and Y, respectively, would you accept the hypothesis that the distributions of X and Y are the same with $\alpha = .05$, using the Wilcoxon-Mann-Whitney test?

5. The total amount of prize money won by the top prize-winning bowler, from 1959 through 1968 was 7.7, 22.5, 26.3, 50.0, 46.3, 33.6, 47.7, 54.7, 54.2, 67.4 (thousands of dollars). Does this look like a random sequence of values from the same population?

6. The *rank correlation coefficient* is frequently used to measure the degree of association between two rankings x_i, y_i for the same objects (thus each of x_i, y_i will take on the values $1, 2, \ldots, n$, in some order, if n items have been ranked). The rank correlation coefficient is defined to be

$$r = \frac{\sum (x_i - \bar{x})(y_i - \bar{y})}{\sqrt{\sum (x_i - \bar{x})^2 \sum (y_i - \bar{y})^2}} = 1 - \frac{6 \sum d_i^2}{n(n^2 - 1)}$$

where $d_i = x_i - y_i$ is the difference in ranks for the same object. Verify this equality.

7. In training inspectors to grade live beef, it is common that a group of, say, 10 steers will be ranked from best to worst (in terms of fullness, depth of fat, etc.) both by a trained, accepted inspector and by a candidate inspector. If the rank correlation between the two is not sufficiently large, the candidate must receive more training before being allowed to inspect on his own. Assume the following rankings were obtained from the inspector (A) and the candidate (B) for 10 steers:

	Steer number									
Ranks by	1	2	3	4	5	6	7	8	9	10
A	9	10	4	1	3	2	6	8	5	7
B	10	8	6	3	2	1	7	9	4	5

Compute the rank correlation coefficient for this data.

8. Another way of testing the randomness of a sequence of values is to form a new sequence from the original, by taking differences between successive elements and retaining only the sign of the difference (+ or −). Thus a sequence of n items generates a sequence of $n - 1$ signs. The run test already described can then be applied to this new sequence, and randomness would be rejected if either too few or too many runs are observed. Apply this test to the data in Exercise 3.

Appendix

TABLE 1. Standard Normal Distribution Function

Entries are values of $z(\alpha)$ such that

$$\alpha = \int_{-\infty}^{z(\alpha)} \frac{1}{\sqrt{2\pi}} e^{-(x^2/2)} dx,$$

for $0 \leq \alpha \leq 3$ in steps of .01

$z(\alpha)$	0	1	2	3	4	5	6	7	8	9
.0	.5000	.5040	.5080	.5120	.5160	.5199	.5239	.5279	.5319	.5359
.1	.5398	.5438	.5478	.5517	.5557	.5596	.5636	.5675	.5714	.5753
.2	.5793	.5832	.5871	.5910	.5948	.5987	.6026	.6064	.6103	.6141
.3	.6179	.6217	.6255	.6293	.6331	.6368	.6406	.6443	.6480	.6517
.4	.6554	.6591	.6628	.6664	.6700	.6736	.6772	.6808	.6844	.6879
.5	.6915	.6950	.6985	.7019	.7054	.7088	.7123	.7157	.7190	.7224
.6	.7257	.7291	.7324	.7357	.7389	.7422	.7454	.7486	.7517	.7549
.7	.7580	.7611	.7642	.7673	.7704	.7734	.7764	.7794	.7823	.7852
.8	.7881	.7910	.7939	.7967	.7995	.8023	.8051	.8078	.8106	.8133
.9	.8159	.8186	.8212	.8238	.8264	.8289	.8315	.8340	.8365	.8389
1.0	.8413	.8438	.8461	.8485	.8508	.8531	.8554	.8577	.8599	.8621
1.1	.8643	.8665	.8686	.8708	.8729	.8749	.8770	.8790	.8810	.8830
1.2	.8849	.8869	.8888	.8907	.8925	.8944	.8962	.8980	.8997	.9015
1.3	.9032	.9049	.9066	.9082	.9099	.9115	.9131	.9147	.9162	.9177
1.4	.9192	.9207	.9222	.9236	.9251	.9265	.9279	.9292	.9306	.9319
1.5	.9332	.9345	.9357	.9370	.9382	.9394	.9406	.9418	.9429	.9441
1.6	.9452	.9463	.9474	.9484	.9495	.9505	.9515	.9525	.9535	.9545
1.7	.9554	.9564	.9573	.9582	.9591	.9599	.9608	.9616	.9625	.9633
1.8	.9641	.9649	.9656	.9664	.9671	.9678	.9686	.9693	.9700	.9706
1.9	.9713	.9719	.9726	.9732	.9738	.9744	.9750	.9756	.9761	.9767

TABLE 1. *(continued)*

$z(\alpha)$	0	1	2	3	4	5	6	7	8	9
2.0	.9772	.9778	.9783	.9788	.9793	.9798	.9803	.9808	.9812	.9817
2.1	.9821	.9826	.9830	.9834	.9838	.9842	.9846	.9850	.9854	.9857
2.2	.9861	.9864	.9868	.9871	.9875	.9878	.9881	.9884	.9887	.9890
2.3	.9893	.9896	.9898	.9901	.9904	.9906	.9909	.9911	.9913	.9916
2.4	.9918	.9920	.9922	.9925	.9927	.9929	.9931	.9932	.9934	.9936
2.5	.9938	.9940	.9941	.9943	.9945	.9946	.9948	.9949	.9951	.9952
2.6	.9953	.9955	.9956	.9957	.9959	.9960	.9961	.9962	.9963	.9964
2.7	.9965	.9966	.9967	.9968	.9969	.9970	.9971	.9972	.9973	.9974
2.8	.9974	.9975	.9976	.9977	.9977	.9978	.9979	.9979	.9980	.9981
2.9	.9981	.9982	.9982	.9983	.9984	.9984	.9985	.9985	.9986	.9986
3.0	.9987									

TABLE 2. Chi-Square Distribution Function

Entries are values of $\chi^2(\alpha)$ such that

$$\alpha = \int_0^{\chi^2(\alpha)} \frac{1}{2^{n/2}\Gamma(n/2)} t^{(n/2)-1} e^{-(t/2)}\, dt$$

for various α and degrees of freedom $n = 1, 2, \ldots, 30$.

	α							
n	.01	.05	.10	.25	.75	.90	.95	.99
1	.000	.004	.016	.102	1.324	2.706	3.843	6.637
2	.020	.103	.211	.575	2.773	4.605	5.992	9.210
3	.115	.352	.584	1.212	4.108	6.251	7.815	11.344
4	.297	.711	1.064	1.923	5.385	7.779	9.488	13.277
5	.554	1.145	1.610	2.674	6.626	9.236	11.070	15.085
6	.872	1.635	2.204	3.454	7.841	10.645	12.592	16.812
7	1.239	2.167	2.833	4.255	9.037	12.017	14.067	18.474
8	1.646	2.733	3.490	5.071	10.219	13.362	15.507	20.090
9	2.088	3.325	4.168	5.899	11.389	14.684	16.919	21.665
10	2.558	3.940	4.865	6.737	12.549	15.987	18.307	23.209
11	3.053	4.575	5.578	7.584	13.701	17.275	19.675	24.724
12	3.571	5.226	6.304	8.438	14.845	18.549	21.026	26.217
13	4.107	5.892	7.041	9.299	15.984	19.812	22.362	27.687
14	4.660	6.571	7.790	10.165	17.117	21.064	23.685	29.141
15	5.229	7.261	8.547	11.036	18.245	22.307	24.996	30.577
16	5.812	7.962	9.312	11.912	19.369	23.542	26.296	32.000
17	6.407	8.682	10.085	12.792	20.489	24.769	27.587	33.408
18	7.015	9.390	10.865	13.675	21.605	25.989	28.869	34.805
19	7.632	10.117	11.651	14.562	22.718	27.203	30.143	36.190
20	8.260	10.851	12.443	15.452	23.828	28.412	31.410	37.566
21	8.897	11.591	13.240	16.344	24.935	29.615	32.670	38.930
22	9.542	12.338	14.042	17.240	26.039	30.813	33.924	40.289
23	10.195	13.090	14.848	18.137	27.141	32.007	35.172	41.637
24	10.856	13.848	15.659	19.037	28.241	33.196	36.415	42.980
25	11.523	14.611	16.473	19.939	29.339	34.381	37.652	44.313
26	12.198	15.379	17.292	20.843	30.435	35.563	38.885	45.642
27	12.878	16.151	18.114	21.749	31.528	36.741	40.113	46.962
28	13.565	16.928	18.939	22.657	32.620	37.916	41.337	48.278
29	14.256	17.708	19.768	23.566	33.711	39.087	42.557	49.586
30	14.954	18.493	20.599	24.478	34.800	40.256	43.773	50.892

TABLE 3. *t* Distribution Function

Entries are values of $t(\alpha)$ such that

$$\alpha = \int_{-\infty}^{t(\alpha)} \frac{\Gamma[(n+1)/2]}{\sqrt{\pi n}\,\Gamma(n/2)} \left[1 + \frac{x^2}{n}\right]^{-(n+1)/2} dx$$

for degrees of freedom $n = 1, 2, 3, \ldots, 30$.

n \ α	.80	.90	.950	.975	.990	.995
1	1.376	3.078	6.134	12.73	31.75	64.00
2	1.061	1.886	2.920	4.303	6.964	9.925
3	.978	1.638	2.353	3.182	4.541	5.841
4	.941	1.533	2.132	2.776	3.747	4.604
5	.920	1.476	2.015	2.571	3.365	4.032
6	.906	1.440	1.943	2.447	3.143	3.707
7	.896	1.415	1.895	2.365	2.998	3.499
8	.889	1.397	1.860	2.306	2.896	3.355
9	.883	1.383	1.833	2.262	2.821	3.250
10	.879	1.372	1.812	2.228	2.764	3.169
11	.876	1.363	1.796	2.201	2.718	3.106
12	.873	1.356	1.782	2.179	2.681	3.054
13	.870	1.350	1.771	2.160	2.650	3.012
14	.868	1.345	1.761	2.145	2.624	2.977
15	.866	1.341	1.753	2.132	2.602	2.947
16	.865	1.337	1.746	2.120	2.584	2.921
17	.863	1.333	1.740	2.110	2.567	2.898
18	.862	1.330	1.734	2.101	2.552	2.878
19	.861	1.328	1.729	2.093	2.540	2.861
20	.860	1.325	1.725	2.086	2.528	2.845
21	.859	1.323	1.721	2.080	2.518	2.831
22	.858	1.321	1.717	2.074	2.508	2.819
23	.858	1.320	1.714	2.069	2.500	2.807
24	.857	1.318	1.711	2.064	2.492	2.797
25	.856	1.316	1.708	2.060	2.485	2.788
26	.856	1.315	1.706	2.056	2.479	2.779
27	.855	1.314	1.703	2.052	2.473	2.771
28	.855	1.312	1.701	2.048	2.467	2.763
29	.854	1.311	1.699	2.045	2.462	2.756
30	.854	1.310	1.697	2.042	2.457	2.750

TABLE 4. *F* Distribution Function

Entries are values of $F_{g,h}(\alpha)$ such that

$$\alpha = \int_0^{F_{g,h}(\alpha)} \frac{\Gamma[(g+h)/2](g/h)^{g/2}}{\Gamma(g/2)\Gamma(h/2)} f^{(g/2)-1}[1+(fg/h)]^{-(g+h)/2}\, df$$

for selected values of g (numerator degrees of freedom), h and α.

α	$h \downarrow$ \ $g \rightarrow$	1	2	3	4	5	6	7	8	9	10	15	20	30	50	100	200
.80	1	9.5	12.0	13.1	13.6	14.0	14.3	14.4	14.6	14.7	14.8	15.0	15.2	15.3	15.4	15.5	15.5
.90	1	39.9	49.5	53.6	55.8	57.2	58.2	58.9	59.4	59.9	60.2	61.2	61.7	62.3	62.7	63.0	63.2
.95	1	161.4	199.5	215.7	224.6	230.2	234.0	236.8	238.9	240.5	241.9	245.9	248.0	250.1	251.8	253.0	253.7
.99	1	4052.2	4999.5	5403.3	5624.5	5763.6	5859.0	5928.3	5981.2	6022.5	6055.9	6157.3	6208.6	6260.7	6302.5	6334.0	6349.9
.80	2	3.5556	4.0000	4.1563	4.2361	4.2844	4.3168	4.3401	4.3576	4.3712	4.3822	4.4151	4.4316	4.4482	4.4614	4.4714	4.4764
.90	2	8.5263	9.0000	9.1618	9.2434	9.2926	9.3255	9.3491	9.3668	9.3805	9.3916	9.4247	9.4413	9.4579	9.4712	9.4812	9.4862
.95	2	18.513	19.000	19.164	19.274	19.297	19.329	19.354	19.371	19.385	19.396	19.429	19.445	19.463	19.476	19.486	19.491
.99	2	98.502	98.999	99.165	99.250	99.299	99.332	99.356	99.375	99.389	99.399	99.432	99.450	99.466	99.478	99.491	99.495
.80	3	2.6822	2.8860	2.9359	2.9555	2.9652	2.9707	2.9741	2.9763	2.9779	2.9791	2.9819	2.9830	2.9838	2.9842	2.9844	2.9845
.90	3	5.5383	5.4624	5.3908	5.3426	5.3092	5.2847	5.2662	5.2517	5.2400	5.2304	5.2003	5.1845	5.1681	5.1546	5.1443	5.1390
.95	3	10.128	9.5521	9.2766	9.1172	9.0135	8.9406	8.8867	8.8452	8.8123	8.7855	8.7029	8.6602	8.6166	8.5810	8.5539	8.5402
.99	3	34.116	30.816	29.457	28.710	28.237	27.911	27.672	27.489	27.345	27.229	26.872	26.690	26.505	26.354	26.240	26.183
.80	4	2.3507	2.4721	2.4847	2.4826	2.4780	2.4733	2.4691	2.4654	2.4623	2.4596	2.4503	2.4450	2.4392	2.4342	2.4302	2.4281
.90	4	4.5448	4.3246	4.1909	4.1072	4.0506	4.0098	3.9790	3.9549	3.9357	3.9199	3.8704	3.8443	3.8174	3.7952	3.7782	3.7695
.95	4	7.7086	6.9443	6.5914	6.3882	6.2561	6.1631	6.0942	6.0410	5.9988	5.9644	5.8578	5.8026	5.7459	5.6995	5.6641	5.6461
.99	4	21.198	18.000	16.694	15.977	15.522	15.207	14.976	14.799	14.659	14.546	14.198	14.020	13.838	13·690	13.577	13.520
.80	5	2.1782	2.2591	2.2530	2.2397	2.2275	2.2174	2.2090	2.2021	2.1963	2.1914	2.1751	2.1660	2.1562	2.1479	2.1413	2.1379
.90	5	4.0604	3.7797	3.6194	3.5202	3.4530	3.4045	3.3679	3.3393	3.3163	3.2974	3.2380	3 2066	3.1741	3.1471	3.1263	3.1157
.95	5	6.6079	5.7861	5.4094	5.1922	5.0503	4.9503	4.8759	4.8183	4.7725	4.7351	4.6188	4.5581	4.4957	4.4444	4.4051	4.3851
.99	5	16.258	13.274	12.060	11.392	10.967	10.672	10.456	10.289	10.158	10.051	9.7222	9.5527	9.3794	9.2378	9.1299	9.0754
.80	6	2.0729	2.1299	2.1126	2.0924	2.0755	2.0619	2.0508	2.0417	2.0342	2.0278	2.0068	1.9951	1.9825	1.9717	1.9632	1.9588
.90	6	3.7760	3.4633	3.2888	3.1808	3.1075	3.0546	3.0145	2.9830	2.9577	2.9369	2.8712	2.8363	2.8000	2.7697	2.7463	2.7343
.95	6	5.9874	5.1433	4.7571	4.5337	4.3874	4.2839	4.2067	4.1468	4.0990	4.0600	3.9381	3.8742	3.8082	3.7537	3.7117	3.6904
.99	6	13.745	10.925	9.7796	9.1483	8.7459	8.4661	8.2600	8.1017	7.9761	7.8741	7.5590	7.3958	7.2285	7.0915	6.9867	6.9336
.80	7	2.0020	2.0434	2.0186	1.9937	1.9736	1.9575	1.9445	1.9339	1.9251	1.9176	1.8930	1.8793	1.8646	1.8519	1.8419	1.8367
.90	7	3.5894	3.2574	3.0741	2.9605	2.8833	2.8274	2.7849	2.7516	2.7247	2.7025	2.6322	2.5947	2.5555	2.5226	2.4971	2.4841
.95	7	5.5915	4.7374	4.3468	4.1203	3.9715	3.8660	3.7870	3.7257	3.6767	3.6365	3.5107	3.4445	3.3758	3.3189	3.2749	3.2525
.99	7	12.246	9.5466	8.4513	7.8466	7.4604	7.1914	6.9928	6.8401	6.7188	6.6200	6.3143	6.1554	5.9920	5.8577	5.7547	5.7024

TABLE 4. *(continued)*

α	h ↓ / g →	1	2	3	4	5	6	7	8	9	10	15	20	30	50	100	200
.80	8	1.9511	1.9814	1.9512	1.9230	1.9005	1.8826	1.8682	1.8564	1.8466	1.8383	1.8109	1.7956	1.7791	1.7648	1.7535	1.7476
.90	8	3.4579	3.1131	2.9238	2.8064	2.7264	2.6683	2.6241	2.5893	2.5612	2.5380	2.4642	2.4246	2.3830	2.3481	2.3208	2.3068
.95	8	5.3177	4.4590	4.0662	3.8379	3.6875	3.5806	3.5005	3.4381	3.3881	3.3472	3.2184	3.1503	3.0794	3.0204	2.9747	2.9513
.99	8	11.259	8.6491	7.5910	7.0061	6.6318	6.3707	6.1776	6.0289	5.9106	5.8143	5.5151	5.3591	5.1981	5.0654	4.9633	4.9114
.80	9	1.9128	1.9349	1.9007	1.8699	1.8455	1.8262	1.8107	1.7979	1.7874	1.7784	1.7488	1.7321	1.7141	1.6985	1.6860	1.6795
.90	9	3.3603	3.0064	2.8129	2.6927	2.6106	2.5509	2.5053	2.4694	2.4403	2.4163	2.3396	2.2983	2.2547	2.2180	2.1892	2.1744
.95	9	5.1174	4.2565	3.8625	3.6331	3.4817	3.3737	3.2927	3.2296	3.1789	3.1373	3.0061	2.9365	2.8637	2.8028	2.7556	2.7313
.99	9	10.561	8.0215	6.9919	6.4221	6.0569	5.8018	5.6129	5.4671	5.3511	5.2565	4.9621	4.8080	4.6486	4.5167	4.4150	4.3631
.80	10	1.8829	1.8986	1.8614	1.8286	1.8027	1.7823	1.7658	1.7523	1.7411	1.7316	1.7000	1.6823	1.6629	1.6461	1.6327	1.6256
.90	10	3.2850	2.9245	2.7277	2.6053	2.5216	2.4606	2.4140	2.3772	2.3473	2.3226	2.2435	2.2007	2.1554	2.1171	2.0869	2.0713
.95	10	4.9646	4.1028	3.7083	3.4780	3.3258	3.2172	3.1355	3.0717	3.0204	2.9782	2.8450	2.7740	2.6996	2.6371	2.5884	2.5634
.99	10	10.044	7.5594	6.5523	5.9943	5.6363	5.3858	5.2001	5.0567	4.9424	4.8492	4.5581	4.4054	4.2469	4.1155	4.0137	3.9617
.80	15	1.7972	1.7952	1.7490	1.7103	1.6801	1.6561	1.6368	1.6209	1.6076	1.5964	1.5584	1.5367	1.5127	1.4914	1.4741	1.4649
.90	15	3.0732	2.6952	2.4898	2.3614	2.2730	2.2081	2.1582	2.1185	2.0862	2.0593	1.9722	1.9243	1.8728	1.8284	1.7929	1.7743
.95	15	4.5431	3.6823	3.2874	3.0556	2.9013	2.7905	2.7066	2.6408	2.5876	2.5437	2.4034	2.3275	2.2468	2.1780	2.1234	2.0950
.99	15	8.6831	6.3589	5.4170	4.8932	4.5556	4.3183	4.1415	4.0045	3.8948	3.8049	3.5222	3.3719	3.2141	3.0814	2.9772	2.9235
.80	20	1.7565	1.7462	1.6958	1.6543	1.6218	1.5960	1.5752	1.5581	1.5436	1.5313	1.4897	1.4656	1.4385	1.4143	1.3941	1.3833
.90	20	2.9747	2.5893	2.3801	2.2489	2.1582	2.0913	2.0397	1.9985	1.9649	1.9367	1.8449	1.7938	1.7382	1.6896	1.6501	1.6292
.95	20	4.3512	3.4928	3.0984	2.8661	2.7109	2.5990	2.5140	2.4471	2.3928	2.3479	2.2033	2.1242	2.0391	1.9656	1.9066	1.8755
.99	20	8.0960	5.8489	4.9382	4.4307	4.1027	3.8714	3.6987	3.5644	3.4567	3.3682	3.0880	2.9377	2.7785	2.6430	2.5353	2.4792
.80	30	1.7172	1.6990	1.6445	1.6001	1.5654	1.5378	1.5154	1.4968	1.4812	1.4678	1.4220	1.3949	1.3641	1.3358	1.3116	1.2983
.90	30	2.8807	2.4887	2.2761	2.1422	2.0492	1.9803	1.9269	1.8841	1.8490	1.8196	1.7223	1.6673	1.6065	1.5522	1.5069	1.4824
.95	30	4.1709	3.3158	2.9223	2.6896	2.5336	2.4205	2.3343	2.2662	2.2107	2.1646	2.0148	1.9317	1.8409	1.7609	1.6950	1.6597
.99	30	7.5625	5.3903	4.5097	4.0179	3.6990	3.4735	3.3045	3.1726	3.0565	2.9791	2.7002	2.5487	2.3860	2.2450	2.1307	2.0700
.80	50	1.6867	1.6624	1.6048	1.5581	1.5216	1.4924	1.4687	1.4490	1.4323	1.4179	1.3682	1.3383	1.3035	1.2706	1.2415	1.2249
.90	50	2.8087	2.4120	2.1967	2.0608	1.9660	1.8954	1.8405	1.7963	1.7598	1.7291	1.6269	1.5681	1.5018	1.4409	1.3885	1.3590
.95	50	4.0343	3.1826	2.7900	2.5572	2.4004	2.2864	2.1992	2.1299	2.0733	2.0261	1.8714	1.7841	1.6872	1.5995	1.5249	1.4835
.99	50	7.1706	5.0566	4.1993	3.7195	3.4077	3.1864	3.0202	2.8900	2.7850	2.6981	2.4190	2.2652	2.0976	1.9490	1.8248	1.7567
.80	100	1.6643	1.6356	1.5757	1.5273	1.4894	1.4591	1.4343	1.4136	1.3961	1.3809	1.3278	1.2954	1.2569	1.2191	1.1839	1.1626
.90	100	2.7564	2.3564	2.1394	2.0019	1.9057	1.8339	1.7778	1.7324	1.6949	1.6632	1.5566	1.4943	1.4227	1.3548	1.2934	1.2571
.95	100	3.9361	3.0873	2.6955	2.4626	2.3053	2.1906	2.1025	2.0323	1.9748	1.9267	1.7675	1.6764	1.5733	1.4772	1.3917	1.3416
.99	100	6.8953	4.8239	3.9837	3.5127	3.2059	2.9877	2.8233	2.6943	2.5898	2.5033	2.2230	2.0666	1.8933	1.7353	1.5977	1.5184
.80	200	1.6533	1.6225	1.5614	1.5122	1.4736	1.4427	1.4173	1.3961	1.3781	1.3625	1.3076	1.2738	1.2329	1.1919	1.1521	1.1266
.90	200	2.7308	2.3293	2.1114	1.9732	1.8763	1.8038	1.7470	1.7011	1.6630	1.6308	1.5218	1.4575	1.3826	1.3100	1.2418	1.1991
.95	200	3.8884	3.0411	2.6497	2.4168	2.2592	2.1441	2.0556	1.9849	1.9269	1.8783	1.7166	1.6233	1.5164	1.4146	1.3206	1.2626
.99	200	6.7633	4.7128	3.8810	3.4143	3.1100	2.8933	2.7298	2.6012	2.4971	2.4106	2.1294	1.9713	1.7941	1.6295	1.4811	1.3912

References

The following books give good discussions of linear algebra and matrices:

1. Perlis, Sam, *Theory of Matrices*, 1952, Addison-Wesley Press, Inc., Cambridge, Mass.

2. Hadley, G., *Linear Algebra*, 1961, Addison-Wesley Publishing Co., Inc., Reading, Mass.

3. Stein, F. Max, *Introduction to Matrices and Determinants*, 1967, Wadsworth Publishing Co., Inc., Belmont, California.

The following books provide good introductions to probability theory:

4. Parzen, E., *Modern Probability Theory and its Applications*, 1960, John Wiley, New York.

5. Barr, Donald R., and Peter W. Zehna, *Probability*, 1971, Wadsworth Publishing Co., Inc., Belmont, California.

This reference develops mathematical statistics from a decision theory viewpoint.

6. Ferguson, Thomas S., *Mathematical Statistics*, 1967, Academic Press, New York.

This reference is a classic in mathematical statistics and presents much more mathematical detail than has the present text.

7. Cramér, Harald, *Mathematical Methods of Statistics*, seventh printing, 1957, Princeton University Press, Princeton, N.J.

Answers To Even-Numbered Problems

Exercise 1.2

2. $B = (A \cap B) \cup (\bar{A} \cap B)$ and $(A \cap B) \cap (\bar{A} \cap B) = \phi$. Thus $P(B) = P(A \cap B) + P(\bar{A} \cap B)$. $A \subset B \Rightarrow A \cap B = A$ so

$$P(B) = P(A) + P(\bar{A} \cap B) \geq P(A) \quad \text{since} \quad P(\bar{A} \cap B) \geq 0.$$

4.

x	2	3	4	5	6	7	8	9	10	11	12
$36p_X(x)$	1	2	3	4	5	6	5	4	3	2	1

6. 4/7

8. .216, .3024

10. .0547

12. 10/11

16. $Q = E[(X - b)^2] = E[X^2] - 2b\mu_X + b^2$

$$dQ/db = 0 \Leftrightarrow b = \mu_X$$

18. $\dfrac{dc_X(t)}{dt} = \dfrac{1}{m_X(t)} m_X'(t), \qquad \dfrac{dc_X(t)}{dt} = \dfrac{1}{[m_X(t)]^2} [m_X''(t) - [m_X'(t)]^2]$

20. $1 - e^{-t}, \qquad t > 0$

22. $\eta_1 = \xi_1, \qquad \eta_2 = \xi_2 + \xi_1, \qquad \eta_3 = \xi_3 + 3\xi_2 + \xi_1$

$\xi_1 = \eta_1, \qquad \xi_2 = \eta_2 - \eta_1, \qquad \xi_3 = \eta_3 - 3\eta_2 + 2\eta_1$

Exercise 1.3

2. Poisson, $\lambda s = 48$, 48

4. 1, 2

6. $1 - q^{[t]}, \qquad t \geq 0$

8. $\dfrac{r-1}{\lambda}$

10. $\dfrac{\ln 2}{\lambda}$, μ

12. no

18. $\dfrac{1}{2\sqrt{b(t-a)}}\left[f_X\left(\sqrt{\dfrac{t-a}{b}}\right) + f_X\left(-\sqrt{\dfrac{t-a}{b}}\right)\right]$

20. $\xi_p = p$, $\xi_p = (b-a)p + a$

22. $-\dfrac{1}{\lambda}\ln(1-p)$

24. Gamma, $\lambda = 1/2$, $r = 1$

Exercise 1.4

2. 545/15,552

12. $f_Y(y) = y \qquad 0 < y < 1$
    $= 1 - y \qquad 1 < y < 2$

18. $P(X > a + b \mid X > b) = P(X > a)$

20. Poisson with $\lambda s = \lambda p$

22. $1 - [1 - F_X(t)]^n$, $[F_X(t)]^n$

Exercise 2.2

10. $N(\mu, (\sigma/\sqrt{n}))$

12. $\dfrac{(n\lambda)^n}{\Gamma(n)} y^{n-1} e^{-n\lambda y}$

14. $\dbinom{mn}{kn} p^{kn}(1-p)^{mn-kn}$, $k = 0, \dfrac{1}{n}, \dfrac{2}{n}, \ldots, m$

16. $MVN(\mu\mathbf{1}, \sigma^2\mathbf{I})$

Exercise 2.3

2. $P\left(\chi_9^2 > \dfrac{n-1}{\sigma^2}\right)$

6. $\chi_2'^2$

12. $\dfrac{h}{h-2}, \dfrac{2h^2[h+g-2]}{g(h-2)^2(h-4)}$

Exercise 3.2

2. $\overline{X}, \overline{X}$

4. $\dfrac{1}{\overline{X}}, \dfrac{1}{\overline{X}}$

6. $\dfrac{1}{\overline{X}}, \dfrac{1}{\overline{X}}$

8. $\overline{X}, \overline{X}$

10. both same, $\overline{X}_1$, $\overline{X}_2$, $\dfrac{n-1}{n} S_1^2$, $\dfrac{n-1}{n} S_2^2$,

$$\frac{\sum (X_{1i}-\overline{X}_1)(X_{2i}-\overline{X}_2)}{\sqrt{\sum (X_{1i}-\overline{X}_1)^2 \sum (X_{2i}-\overline{X}_2)^2}}$$

12. $\overline{X}, \dfrac{1}{2}\sqrt{1+\dfrac{4}{n}\sum X_i^2} - \dfrac{1}{2}$

14. $1 - e^{-(1/3)}$

18. $\dfrac{1}{5}(\overline{X}_1 + \overline{X}_2)$

20. $\hat{\alpha} = \dfrac{N_Z^{-1}(f_1)\log c_2 - N_Z^{-1}(f_2)\log c_1}{\log c_2 - \log c_1}, \ \hat{\beta} = \dfrac{N_Z^{-1}(f_2) - N_Z^{-1}(f_1)}{\log c_2 - \log c_1}$

26. $nt, \dfrac{nT_0}{k}$

28. $\sqrt{\dfrac{3}{n}\sum x_i^2}$, $\max |x_i|$

30. $\overline{X} = (n_1\overline{X}_1 + n_2\overline{X}_2)/(n_1+n_2)$, $\sum_j (\overline{X}_{1j} - \overline{X})^2/n_1$, $\sum_j (\overline{X}_{2j} - \overline{X})^2/n_2$

32. $\hat{\alpha} = -5.8$, $\hat{\beta} = 1$, $\hat{\mu} = 5.8$, $\hat{\sigma} = 1$

Exercise 3.3

2. all $\sum X_i$

18. $n_i/(\sum n_j)$

20. $N(\mu, \sigma/\sqrt{n})$, $N\left(\frac{n-1}{n}\sigma^2, \frac{\sigma^2}{n}\sqrt{2(n-1)}\right)$, $N(\sigma^2, \sigma^2\sqrt{2/(n-1)})$

22. $N(\lambda, \sqrt{\lambda/n})$

Exercise 4.1

4. $1 - \alpha$

Exercise 4.2

2. Minimize interval length

4. 32

6. $172.13\sigma^2$

8. $\bar{X}_A + \bar{X}_B + S\sqrt{\frac{2}{n}}\,t_{2(n-1)}(1 - (\alpha/2))$,

   $S^2 = [\sum (X_{Ai} - \bar{X}_A)^2 + \sum (X_{Bi} - \bar{X}_B)^2]/2(n-1)$

10. $(S\sqrt{(n-1)/\chi^2(1-(\alpha/2))}, S\sqrt{(n-1)/\chi^2(\alpha/2)})$

14. $t_k = \mu + \sigma z_k$, $(\min_R(\mu + \sigma z_k), \max_R(\mu + \sigma z_k))$, R is region of Theorem 4.2.5

16. $(X_{(n)}/\sqrt[n]{1 - (\alpha/2)}, X_{(n)}/\sqrt[n]{\alpha/2})$

18. All $\boldsymbol{\mu}$ in region $\frac{n(n-p)}{p}(\bar{\mathbf{x}} - \boldsymbol{\mu})'\mathbf{s}^{-1}(\bar{\mathbf{x}} - \boldsymbol{\mu}) \le F(1-\alpha)$

22. $(\sum X_i^2/\chi_n^2(1 - (\alpha/2)), \sum X_i^2/\chi_n^2(\alpha/2))$, shorter expected length

Exercise 4.3

4. same, constant

6. $\frac{Y}{n} \pm z(1 - (\alpha/2))\sqrt{\frac{Y(n-Y)}{n^3}}$

8. $\frac{r}{\bar{x}} \pm z(1 - (\alpha/2)) \frac{1}{\bar{x}} \sqrt{\frac{r(\bar{x} - r)}{n\bar{x}}}, \ \bar{x} \pm z(1 - (\alpha/2))\bar{x} \sqrt{\frac{\bar{x} - r}{n\bar{x}r}}$

10. $S^2 \pm z(1 - (\alpha/2))S^2 \sqrt{\frac{2}{n}}$

Exercise 5.1

2. (*b*)

4. .01

Exercise 5.2

2. $c = .58, n = 16$

4. $\bar{x} \leq c$, see binomial tables

6. $\sum x_i^2(\sigma_0^{-2} - \sigma_1^{-2}) - 2n\bar{x}(\mu_0\sigma_0^{-2} - \mu_1\sigma_1^{-2}) < c$

8. $\bar{X} > \frac{\chi^2(1 - \alpha)}{2\lambda_0 n}$, $\bar{X} < 156{,}219$, more than .995

10. $\prod X_i^2 < c$

12. $\prod X_i(1 - X_i) < c$

14. $\sum X_i < c$

16. $\bar{X}_1 - \bar{X}_2 > -z(1 - \alpha)\sqrt{(2/n)}$

Exercise 5.3

2. Yes

4. $S^2 > c_1$ if $\sigma_1^2 > \sigma_0^2$, no

8. No

12. $\mu_1 - \mu_2$, $\sigma_{11} + \sigma_{22} - 2\sigma_{12}$, reject if $|\bar{Y}| > t(1 - (\alpha/2))S_Y/\sqrt{n}$

Exercise 5.4

2. Yes, $\alpha = .1$

6. $[(\prod X_{i.}^{X_{i.}})(\prod X_{.j}^{X_{.j}})]/\prod X_{ij}^{X_{ij}} < c$

10. $U_i = F_{W_i}(W_i)$, $\chi_{2k}^2 = \sum_i - 2 \ln U_i$

Exercise 6.1

4. $p \geq 3/4$

6. $r(3 - 4r^2)/6n$

Exercise 6.2

8. $(n + 1)/(\sum X_i + 2)$

10. Approaches $\bar{X}$, approaches μ_0

Exercise 6.3

2. $(\theta \ln |x|^{-1})^{-1}$ for $1/2 < |x| < 1$, $|x| < \theta < 1$
   $(\theta \ln 2)^{-1}$ for $|x| < 1/2$, $1/2 < \theta < 1$
   $(|x|, |x|^{\alpha})$ if $1/2 < |x| < 1$, $(1/2, 1/2^{\alpha})$ if $|x| < 1/2$

4. $\bar{x} < 9.79$, $\alpha = .00000$, $\beta = .9973$

6. Accept $p = p_1 \Leftrightarrow (p_2/p_1)^x < [(1 - p_1)/(1 - p_2)]^{n-x}$

Exercise 7.1

2. $\text{tr}(\mathbf{A}) = \text{tr}(\mathbf{CC'A}) = \text{tr}(\mathbf{C'AC}) = \text{tr}(\mathbf{D}) = \sum \lambda_i = r(\mathbf{A})$

4. $\mathbf{A}^{-1}(\mathbf{A}^2) = \mathbf{A}^{-1}(\mathbf{A}) = \mathbf{I}$

8. Multiply the matrices of the two forms

Exercise 7.2

2. $\hat{A} = \bar{Y} - \hat{B}\bar{x} \Rightarrow \hat{A} + \hat{B}\bar{x} = \bar{Y}$
   $\mathbf{1x(x'x)^{-1}x'} = \mathbf{1}$ since $x_1 \equiv 1$

8. (a) $(1/n) \sum Y_i/x_i$ (b) $\bar{Y}/\bar{x}$

10. $[\sum 1/x_i]/n^2 \leq n/(\sum \sqrt{x_i})^2$

12. Remainder decreases with n

Exercise 7.3

2. $((n - 2)S^2/\chi_6^2(1 - (\alpha/2)), (n - 2)S^2/\chi_6^2(\alpha/2))$, yes

6. $F(2, 7) = 1.81$, accept H_1

8. $F(1, 7) = 1.02$, accept H_0

10. No

Exercise 7.4

8. $F(3, 11) = 12.15$, No

Exercise 8.2

2. $P(X_{(1)} = 1, X_{(2)} = 2) = P(X_{(1)} = 1, X_{(2)} = 3)$
   $= P(X_{(1)} = 2, X_{(2)} = 3) = 1/3$
   $P(X_{(1)} = 1) = 2P(X_{(1)} = 2) = 2/3, 2P(X_{(2)} = 2)$
   $= P(X_{(2)} = 3) = 2/3$

8. $6(e^{-2x} - e^{-3x}), F_X(E[X_{(2)}]) = 1 - e^{-5/6} = .564$
   $E[F_X(X_{(2)})] = 1/2$

9. $n = 199$

10. .5858

12. $X_{((n+1)/2)} - \theta + (1/2)$ is beta, $\alpha = \beta = (n - 1)/2$

Exercise 8.3

2. 94 cases out of 126 lead to an absolute mean difference of 7.35 or more. Accept equality.

4. $u = 177, z = -1.28$, accept H_0

8. Accept.

Index

Applied Probability and Statistics (Continued)

WOLD and JUREEN · Demand Analysis

WONNACOTT and WONNACOTT · Introduction to Econometric Methods

YOUDEN · Statistical Methods for Chemists

ZELLNER · An Introduction to Bayesian Inference in Econometrics

Tracts on Probability and Statistics

BILLINGSLEY · Ergodic Theory and Information

BILLINGSLEY · Convergence of Probability Measures

CRAMÉR and LEADBETTER · Stationary and Related Stochastic Processes

JARDINE and SIBSON · Mathematical Taxonomy

KINGMAN · Regenerative Phenomena

RIORDAN · Combinatorial Identities

TAKÁCS · Combinatorial Methods in the Theory of Stochastic Processes